Preparation and Application of Cellulose-Based and Chitosan-Based Materials

Preparation and Application of Cellulose-Based and Chitosan-Based Materials

Guest Editors

Guangmei Xia
Peng Jia

MDPI

Basel • Beijing • Wuhan • Barcelona • Belgrade • Novi Sad • Cluj • Manchester

Guest Editors

Guangmei Xia
State Key Laboratory of Biobased Material and Green Papermaking
Qilu University of Technology
Jinan
China

Peng Jia
State Key Laboratory of Biobased Material and Green Papermaking
Qilu University of Technology
Jinan
China

Editorial Office
MDPI AG
Grosspeteranlage 5
4052 Basel, Switzerland

This is a reprint of the Special Issue, published open access by the journal *Polymers* (ISSN 2073-4360), freely accessible at: https://www.mdpi.com/journal/polymers/special_issues/DW6EIQ9Y67.

For citation purposes, cite each article independently as indicated on the article page online and as indicated below:

Lastname, A.A.; Lastname, B.B. Article Title. *Journal Name* **Year**, *Volume Number*, Page Range.

ISBN 978-3-7258-7959-5 (Hbk)
ISBN 978-3-7258-7960-1 (PDF)
https://doi.org/10.3390/books978-3-7258-7960-1

Contents

About the Editors

Guangmei Xia

Guangmei Xia obtained her Ph.D. from the Institute of Chemistry, Chinese Academy of Sciences. She is currently a faculty member at the School of Light Industry, Qilu University of Technology. Her primary research interests lie in biomass pretreatment and high-value utilization, polymer composites, and bio-based functional materials. She has led several provincial and ministerial-level projects, including the General Program of the China Postdoctoral Science Foundation, the Shandong Provincial Natural Science Foundation, and the Shandong Postdoctoral Innovation Project, and has participated in major national and provincial key projects such as the National Natural Science Foundation of China, the Shandong Major Scientific and Technological Innovation Project, the National Key R&D Program of China, and the Shandong Youth Innovation Team. She has published over 40 peer-reviewed academic papers and been granted 6 domestic and foreign invention patents. Internationally, she is actively engaged in the academic community and is invited to serve as a peer reviewer for top-tier journals including *Journal of Environmental Chemical Engineering, Polymers, Nanomaterials, Gels, Carbohydrate Polymers, ACS Applied Materials & Interfaces, Biomass Conversion and Biorefinery*, and *Cellulose*, among others. Beyond her research, she has supervised undergraduate students to win more than 30 provincial and above awards, and has been honored as the Gold Medal Instructor of the Shandong College Students' Science and Technology Innovation Competition and the Excellent Instructor of the Shandong College Students' Extracurricular Academic Science and Technology Works Competition. She also serves as a reviewer for academic dissertations and Shandong College Students' Innovation Projects, won the Gold Award in the Challenge and Lead Project Competition of the China•Shandong Postdoctoral Innovation and Entrepreneurship Competition, and is a holder of the Shandong Huicai Card.

Peng Jia

Peng Jia received a BS degree in Material Chemistry from Xi'an University of Technology in 2011 and an MS degree and a PhD degree in Materials Science and Engineering from the University of Jinan in 2015 and 2019. He has worked as an assistant professor in Prof. Xingxiang Ji's group, State Key Laboratory of Green Papermaking and Resource Recycling/MOE Key Laboratory of Pulp and Paper Science & Technology, Qilu University of Technology (Shandong Academy of Sciences) since 2019. His research interests include the design and synthesis of hybrid cellulose-based nanocomposites toward highly efficient catalysis, energy storage, sensing, and environmental remediation. He has presided over and participated in 7 national-level scientific research projects and 9 provincial and ministerial-level scientific research projects; published 36 SCI papers as the first author and corresponding author in international journals such as *SusMat, Applied Catalysis B: Environmental, Polymers*, and *Nanomaterials*; and has been granted 5 Chinese invention patents and 4 international invention patents as the first inventor.

polymers

Editorial

Preparation and Application of Cellulose-Based and Chitosan-Based Materials

Guangmei Xia * and Peng Jia

Key Laboratory of Pulp and Paper Science & Technology of Ministry of Education, State Key Laboratory of Green Papermaking and Resource Recycling, Faculty of Light Industry, Qilu University of Technology (Shandong Academy of Sciences), Jinan 250353, China
* Correspondence: gmxia@qlu.edu.cn

1. Introduction

The distinct inherent features of biomass resources include their carbon neutrality, noncompetition with food crops, abundant reserves, widespread distribution, low cost, and environmental friendliness [1–3]. The conversion of biomass into high-value functional materials, clean energy, and fine chemicals is a highly promising strategy to address pressing global challenges, including environmental degradation, energy scarcity, and climate change [4–6].

Among the diverse structural components of biomass, cellulose is the most abundant natural polymer on Earth, with its extensive sources including wood, agricultural straw, herbaceous plants, microorganisms, and even certain animal tissues. Cellulose can be processed into a wide variety of regenerated cellulose materials, such as cellulose films, cellulose fibers, cellulose sponges, and cellulose hydrogels [7–9]. Furthermore, targeted chemical modification of cellulose enables the synthesis of versatile cellulose derivatives, such as cellulose esters, cellulose ethers, and graft copolymers, which considerably broaden the application scope of cellulose to include numerous industrial and scientific fields [9–11].

Chitin is another critical natural polysaccharide that is extensively distributed in the exoskeletons of crustaceans (e.g., shrimps and crabs) as well as insects. Via deacetylation under alkaline conditions, chitin can be efficiently converted into chitosan. In contrast to insoluble chitin, chitosan exhibits solubility in dilute acidic aqueous solutions, which enhances its processability and facilitates subsequent material fabrication. Chitosan can be constructed into diverse functional materials, including films, fibers, separation membranes, tissue engineering scaffolds, and composite materials owing to its exceptional biodegradability, biocompatibility, film-forming capacity, and excellent adsorption performance. Chitosan thus demonstrates application potential in biomedical engineering, environmental remediation, food packaging, agricultural production, and other high-value sectors [12–14].

This Special Issue, entitled "Preparation and Application of Cellulose-Based and Chitosan-Based Materials", focuses on cutting-edge research advances and novel findings related to cellulose- and chitosan-based materials, covering physical and chemical modification mechanisms, optimized preparation and processing technologies, performance enhancement strategies, and the expansion of novel application fields. This Special Issue collects eight high-quality original research papers, and a brief overview of their core content is provided below to arouse reader interest and encourage in-depth reading of these valuable studies.

2. Overview of Research Results

As illustrated in Figure 1, the eight articles published in this Special Issue are systematically categorized into five distinct groups based on the application fields of the developed cellulose- and chitosan-based materials, namely, the mechanical performance optimization of composites, the development of biomedical functional materials, agricultural biostimulants, environmental pollutant treatment, and food packaging material innovation. This classification demonstrates the vibrant research and broad application prospects in this interdisciplinary field.

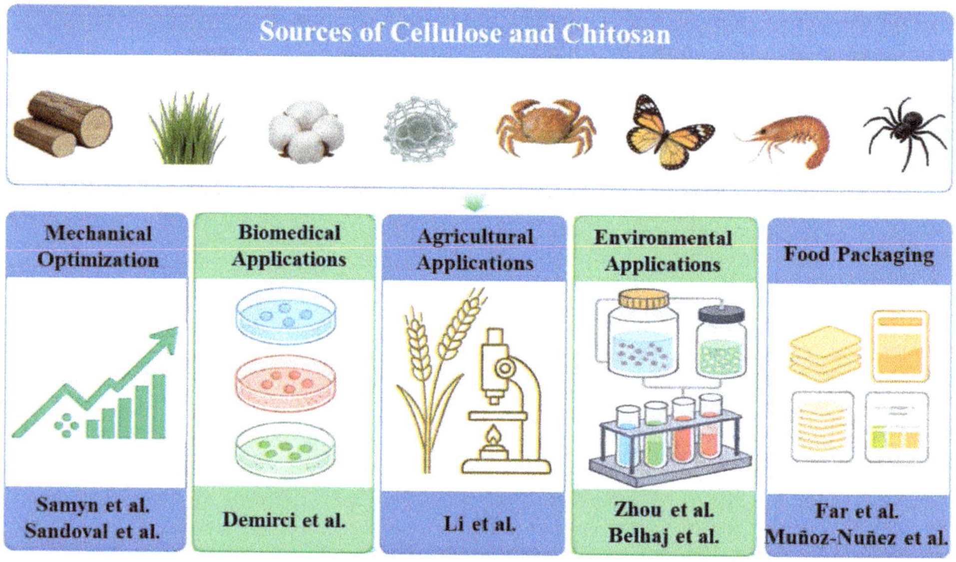

Figure 1. Schematic representation of the research topics covered by articles 1–8 published in this Special Issue, "Preparation and Application of Cellulose-Based and Chitosan-Based Materials", based on the contributor list.

2.1. *Mechanical Performance Optimization of Cellulose-Based Composites*

2.1.1. Cellulose Acetate Microfiber Reinforced Thermoplastic Matrix

Sandoval et al. focused on the green enhancement in the mechanical properties of thermoplastic composites. Low-density polyethylene (LDPE) was selected as the polymer matrix, which was blended with potato starch (PS) to increase biodegradability. Cellulose acetate microfibers (MFCAs) were fabricated via electrospinning and incorporated into the matrix at three gradient loading levels of 0.05%, 0.15%, and 0.30%. The composite blends were processed via extrusion granulation and injection molding to prepare standard test specimens in accordance with ASTM D638 specifications.

The experimental results revealed that the cellulose acetate microfibers effectively reinforced the composite materials owing to their high aspect ratio and interfacial compatibility with the matrix. The tensile strength of the T3 group (84.7% LDPE, 15% PS) loaded with 0.30% MFCA reached 12.32 ± 0.5 MPa, representing an 11.79% increase compared with that of the T4 control group (85% LDPE, 15% PS), to which microfibers were not added. The melt flow index of the composite was precisely regulated, and the melt flow rate of the T2 group (84.85% LDPE, 15% PS) was close to that of pure LDPE, making the T2 group highly suitable for industrial extrusion processing. The study provides a feasible green modification scheme for thermoplastic matrices, and the as-prepared composites show potential for replacing partial synthetic polymers in industrial product manufacturing.

https://doi.org/10.3390/polym18070812

2.1.2. Crosslinked Sprayed Cellulose Nanofibril Coatings

Samyn et al. addressed the critical limitation caused by the insufficient mechanical properties of pristine sprayed cellulose nanofibril (CNF) coatings, which restrict their practical industrial applications. Boric acid (BA) was employed as an efficient internal crosslinker, and polydopamine (PDA) served as an adhesion-promoting layer for fabricating two types of CNF-based coatings on glass substrates: multilayer structured coatings (CNF/BA) and homogeneous mixed coatings (CNF + BA).

The comprehensive performance of the coatings was evaluated via standardized tests including tape adhesion, rub resistance, cross-cut, and scratch resistance tests. The results demonstrated that the PDA/BA adhesion layer (with a BA concentration of 10 mM) endowed the coatings with high interfacial adhesion strength. The scratch resistance of the mixed coating (CNF + 10 mM BA) was 9 N, whereas that of the multilayer coating (CNF/100 mM BA) was 8 N, both of which were substantially higher than the scratch resistance of pristine CNF coatings (<0.5 N). After boric acid crosslinking, the coatings exhibited more stable water contact angles and reduced water penetration rates, alongside increased coating density and a slight decrease in optical transparency. The researchers effectively increased the mechanical durability of sprayed CNF coatings, enabling their practical application in industrial surface protection and high-contact surface protective coatings.

2.2. *Chitosan-Based Materials in Biomedical Applications*

Demirci aimed to overcome the core drawbacks of conventional chitosan surgical sutures, including their insufficient mechanical strength and single functional performance. Novel curcumin-loaded and TiO_2-reinforced chitosan monofilament sutures were successfully fabricated via a dry jet-wet spinning technique.

The systematic characterization of the sutures indicated that adding 3% TiO_2 significantly enhanced the tensile strength of chitosan sutures by 12.32%, reaching 189.41 MPa, while maintaining flexibility and ductility. After curcumin loading, the suture exhibited a free radical scavenging capacity of 43% at 125 h, far exceeding that of pristine chitosan sutures. The 24 h bacterial inactivation rate against *Staphylococcus aureus* was 98.87%, and the cumulative curcumin release rate was 77% within 25 h. Additionally, the suture induced the formation of a hydroxyapatite layer in simulated body fluid, demonstrating promising tissue regeneration potential. This multifunctional suture integrates wound suturing, antibacterial activity, antioxidant capacity, and tissue repair promotion features, indicating this suture is a competitive candidate for absorbable surgical sutures in clinical practice.

2.3. *Chitooligosaccharides in Agricultural Applications*

Li et al. focused on alleviating the adverse impacts of salt stress on agricultural crop production, a major constraint limiting global agricultural yield. Three deacetylases (NodB, VcCOD, and ArCE4A) were utilized to prepare three chitooligosaccharides (N-COS, C-COS, and A-COS) with distinct structural sequence arrangements using chitooligosaccharides with a polymerization degree of 2~6 as raw materials. Their regulatory effects on inducing salt resistance in wheat seedlings were systematically investigated.

All three chitooligosaccharides alleviated salt stress damage to varying degrees, and A-COS had the highest activity. When wheat seedlings under salt stress for 3 days were treated with 10 mg/L A-COS, the leaf length was restored, and the fresh and dry weights were restored by 20.40% and 6.64%, respectively. At the same time, 10 mg/L A-COS promoted proline accumulation and reduced the malondialdehyde content by 34.75%. Moreover, the Na^+/K^+ ratio was significantly reduced. As a result, the ion imbalance and oxidative damage caused by salt stress were effectively alleviated. The authors clarified the effect of the chitooligosaccharide sequence on the induction of plant salt resistance,

https://doi.org/10.3390/polym18070812

providing a theoretical basis for the development of high-efficiency chitooligosaccharide-based plant biostimulants.

2.4. Cellulose- and Chitosan-Based Materials in Environmental Applications

2.4.1. Phosphonated Chitosan for Separating Rare Earth Ions

Zhou et al. targeted key challenges: the difficulty of separating rare earth ions and the low stability of chelating agents under acidic conditions, which hinder the efficient recovery of rare earth resources. In the work, acidic phosphonic chitosan (aPCS) was successfully synthesized via chemically modifying chitosan through the Mannich reaction.

Density functional theory (DFT) calculations revealed that the phosphonic acid groups of aPCS serve as the preferred binding sites for rare earth ions by forming stable bidentate chelate complexes. Complexation–'ultrafiltration experiments demonstrated that, at pH = 5 and a P/RE molar ratio of 10, the rejection rate of aPCS for La(III) was 97%, far exceeding the 70% rejection rate of conventional phosphorylated chitosan (PCS). Across the acidic pH range of 3~7, aPCS consistently exhibited superior La(III) rejection performance than PCS. This functional material holds promise for the selective extraction and purification of rare earth ions in acidic aqueous solutions, providing a novel and efficient solution for sustainably recovering rare earth resources.

2.4.2. Agricultural-Waste-Derived Cellulose/ZnO Composites for Dye Removal

Belhaj et al. utilized agricultural pruning waste (almond and fig tree branches) as a low-cost and sustainable raw material for cellulose extraction. Cellulose/ZnO composites were prepared via the in situ deposition of ZnO nanoparticles for treating textile dye wastewater.

The characterization results showed that the size of the ZnO nanoparticles in the composite ranged from 13 to 26 nm, which were significantly smaller than pure ZnO nanoparticles (43 ± 12 nm). The ZnO nanoparticles were uniformly dispersed on the cellulose matrix. The composite removed 65% of the methylene blue (cationic dye) within 20 min and 45% of the anionic dyes including methyl orange and bromophenol blue. After three repeated adsorption–photocatalysis cycles, the removal rate for bromophenol blue only decreased by 15–19%, confirming their high reusability and stability. The researchers achieved the high-value utilization of agricultural waste biomass, providing a sustainable dual-functional adsorption–photocatalytic material for efficient textile wastewater treatment.

2.5. Cellulose- and Chitosan-Based Materials in Food Packaging Applications

2.5.1. Air-Assisted Sprayed Cellulose Acetate/Chitosan Films

Far et al. employed the air-assisted solution spraying (AASS) technique to fabricate cellulose acetate (CA)/chitosan (CS) composite films, which were highly flexible, and the addition of external plasticizers was not required.

The optimized composite films exhibited a Young's modulus below 1 GPa, with tensile strength and elongation at break exceeding 19 MPa and 2%, respectively, meeting the mechanical requirements for food packaging materials. With increases in chitosan content, the water and oil contact angles of the films increased, and the water barrier performance significantly improved. When the chitosan content was 7.5%, the film exhibited strong antibacterial activity against *Escherichia coli*, and the bacterial colony count reduced by two orders of magnitude compared with that achieved with neat cellulose acetate films. Compared with the traditional solution casting method, the AASS process features a faster drying speed and lower energy consumption, making the process more environmentally friendly and industrially scalable. The as-prepared sustainable and functional composite films are highly suitable for eco-friendly food packaging applications.

https://doi.org/10.3390/polym18070812

2.5.2. 1-Methylimidazolium-Chitosan-Modified Films for Antimicrobial and Antioxidant Packaging

Muñoz-Nuñez et al. synthesized 1-methylimidazolium-chitosan (CS-MeIm) via chemical modification and incorporated CS-MeIm into chitosan film matrices at a loading of 10 wt%. Additionally, 1 or 5 wt% chitin nanowhiskers (ChNws) were introduced as reinforcing agents to further optimize the film's performance.

Antioxidant capacity of the modified film drastically increased from 3 μmol Trolox per gram of film to 15 μmol Trolox per gram of film. The bacterial inactivation rate against *Escherichia coli* was 99.999%, indicating a significant improvement in antibacterial performance. Meanwhile, the water vapor transmission rate of the film decreased, and the mechanical properties were effectively regulated. Particularly, Young's modulus increased, while the elongation at break slightly decreased. In summary, this modified chitosan film can be used as a biodegradable active packaging material for food preservation, effectively extending the shelf life of perishable foods.

The eight high-quality research papers included in this Special Issue focus on the targeted modification, performance optimization, and diversifying the applications of cellulose- and chitosan-based materials, covering core practical demands across multiple key fields:

Mechanical performance optimization: Two effective strategies, namely, the cellulose microfiber reinforcement of thermoplastic matrices and boric acid crosslinking of cellulose nanofibril coatings, significantly increased the mechanical robustness of cellulose-based materials, expanding their industrial application scenarios and practical usability.

Biomedical engineering: The developed TiO_2-reinforced and curcumin-loaded chitosan surgical sutures integrate mechanical properties and multifunctional biological activity, opening a new direction for innovation in absorbable clinical surgical sutures.

Agricultural science: The correlation between chitooligosaccharide's structural sequence and plant salt resistance induction was elucidated, providing reliable theoretical support for the development of green and high-efficiency plant biostimulants to address soil salinization issues.

Environmental remediation: The synthesized acidic phosphonated chitosan enables the efficient separation of rare earth ions in acidic solutions, and agricultural waste-derived cellulose/ZnO composites show synergistic adsorption–photocatalysis effects in textile dye wastewater treatment, enabling resource recovery and environmental pollution control.

Food packaging: Two novel functional chitosan/cellulose films enable plasticizer-free flexible processing and enhanced antibacterial-antioxidant performance, providing feasible and sustainable solutions for the development of eco-friendly food packaging materials.

Collectively, the research findings presented in this Special Issue demonstrate the immense application potential and sustainable value of cellulose- and chitosan-based materials, laying theoretical and experimental foundations for subsequent in-depth research and industrial translation in this field. The growing global research and continuous academic contributions from scholars worldwide have highlighted the urgent need for a follow-up Special Issue in this field, which will be entitled Preparation and Application of Cellulose-Based and Chitosan-Based Materials II.

To conclude this Editorial, I would like to extend a sincere and warm invitation to all researchers dedicated to this dynamic field to contribute their innovative, cutting-edge findings and join us in further advancing the fundamental research and practical application of cellulose- and chitosan-based biomass materials, driving the sustainable development of this promising research domain.

Conflicts of Interest: The authors declare no conflicts of interest.

https://doi.org/10.3390/polym18070812

List of Contributions:

1. Sandoval, P.R.; Rubiano-Navarrete, A.F.; Gómez-Pachón, E.Y.; Vera-Graziano, R. Evaluation of Mechanical Properties of Composite Material with a Thermoplastic Matrix Reinforced with Cellulose Acetate Microfibers. *Polymers* **2024**, *16*, 2557. https://doi.org/10.3390/polym16182557.
2. Samyn, P.; Cosemans, P.; Van der Eycken, E. V.; Coppola, G.A. Enhanced Mechanical Robustness of Sprayed Cellulose Nanofibril Coatings Through Internal Crosslinking with Boric Acid. *Polymers* **2025**, *17*, 2451. https://doi.org/10.3390/polym17182451.
3. Demirci, F. Development of Curcumin-Loaded TiO_2-Reinforced Chitosan Monofilaments for Biocompatible Surgical Sutures. *Polymers* **2025**, *17*, 0484. https://doi.org/10.3390/polym17040484.
4. Li, J.; Li, A.; Li, Y.; Zhu, S.; Song, L.; Liu, S.; Xing, R.; Li, K. Preparation of Chitooligosaccharides with Specific Sequence Arrangement and Their Effect on Inducing Salt Resistance in Wheat Seedlings. *Polymers* **2025**, *17*, 1194. https://doi.org/10.3390/polym17091194.
5. Zhou, M.; Liu, Z.; Lu, D.; Wang, J.; Chen, Z.; Qiu, Y. Synthesis of Acidic Phosphonic Chitosan and the Complexation of La(III) in Acidic Aqueous Solution. *Polymers* **2025**, *17*, 1341. https://doi.org/10.3390/polym17101341.
6. Belhaj, J.; Khiari, R.; Garcia-Caballero, V.; Romero, A.A.; Garcia, A. Agricultural Waste-Derived Cellulose/ZnO Composites: Dual Photocatalytic and Adsorptive Action for Textile Dye Removal. *Polymers* **2025**, *17*, 1737. https://doi.org/10.3390/polym17131737.
7. Far, N.M.; Kramar, A.; Gonzalez-Benito, J. Air-Assisted Sprayed Flexible Cellulose Acetate/Chitosan Materials for Food Packaging. *Polymers* **2025**, *17*, 2479. https://doi.org/10.3390/polym17182479.
8. Muñoz-Nuñez, C.; Quiroz-Pereira, Y.; Muñoz-Bonilla, A.; Fernández-García, M. Enhancing Antimicrobial and Antioxidant Properties of Chitosan-Based Films with 1-Methylimidazolium-Chitosan. *Polymers* **2025**, *17*, 2608. https://doi.org/10.3390/polym17192608.

References

1. Xia, G.; Ma, Y.; Ma, Q.; Yao, X.; Xu, Z.; Ji, X.; Zhang, F. Anti-ultraviolet, antioxidant and pH-responsive cellulose-based composite film incorporated with alizarin for intelligent packaging applications. *Food Packag. Shelf Life* **2024**, *46*, 101413. [CrossRef]
2. Jiang, J.; Wang, Y.; Li, Y.; Huang, J.; Zhang, W.; Wang, D.; Xu, D.; Fu, D.; Zhang, P.; Wang, L.; et al. Solar-driven catalytic platforms for closed-loop carbon cycling: CO_2 assisting biomass conversion. *Bioresour. Technol.* **2026**, *443*, 133789. [CrossRef] [PubMed]
3. Ma, C.; Hu, J.; Wang, H.; Yu, Y.; Tan, C. Advances and challenges in biomass thermochemical conversion: From resource utilization to process optimization. *Renew. Sustain. Energy Rev.* **2026**, *226*, 116385. [CrossRef]
4. Peng, X.; Fu, N.; Lei, Y.; Li, Z.; Cao, M.; Wang, J.; Chen, Y.; Qin, Y.; Chang, Z.; Li, H.; et al. High value utilization of biomass: Moving towards a carbon neutrality future. *Renew. Sustain. Energy Rev.* **2026**, *226*, 116477. [CrossRef]
5. Gong, Z.; Yang, G.; Wu, T.; Wang, M.; Yang, J.; Lei, Z.; Shuai, L. Screening high-quality lignins extracted from biomass to prepare biobased wood adhesives. *Nat. Protoc.* **2026**. [CrossRef] [PubMed]
6. Xu, Y.C.; Zou, S.L.; Wang, Q.; Xiao, L.P. Lignin upgrading for sustainable materials and chemicals: From waste to functional bioplastics. *Bioresour. Technol.* **2026**, *445*, 134091. [CrossRef] [PubMed]
7. Zhang, Y.; Shen, Y.; Liu, R.; Zhang, W.; Luo, Z.; Wang, X.; Chen, Z.; Yu, J.; Wang, X. Research progress of cellulose composite membranes in advanced functional materials: Strategy, performance, and application. *J. Environ. Chem. Eng.* **2026**, *14*, 1107. [CrossRef]
8. Tosif, M.M.; Zannou, O.; Purkiewicz, A.; Chawla, P.; Goksen, G. Recent advances in cellulose-based aerogels for meat packaging: Fabrication, modifications, functionality, and preservation mechanisms. *Trends Food Sci. Technol.* **2026**, *168*, 105511. [CrossRef]
9. Xia, G.; Zhou, Q.; Xu, Z.; Zhang, J.; Zhang, J.; Wang, J.; You, J.; Wang, Y.; Nawaz, H. Transparent cellulose/aramid nanofibers films with improved mechanical and ultraviolet shielding performance from waste cotton textiles by in-situ fabrication. *Carbohydr. Polym.* **2021**, *273*, 118569. [CrossRef] [PubMed]
10. Cui, L.; Shi, X.; Xu, C.; Wu, S. Recent advances in chitosan-based mixed matrix membranes for efficient wastewater treatment. *Int. J. Biol. Macromol.* **2026**, *345*, 150628. [CrossRef] [PubMed]
11. Peng, J.; Ji, X.; Xia, X.; Li, C.; Wei, Z.; Chu, C.; Xu, Z.; Zhang, J.; Xia, G. Cellulose-Based Light-Management Films with Improved Properties Directly Fabricated from Green Tea. *Polysaccharides* **2022**, *3*, 776–791. [CrossRef]
12. Khan, O.; Bhawale, R.; Vasave, R.; Mehra, N.K. Chitosan conjugates in precision medicine: From synthesis to their role in advanced drug delivery. *Int. J. Biol. Macromol.* **2026**, *350*, 151010. [CrossRef]

https://doi.org/10.3390/polym18070812

13. Huang, Z.; Tian, S.; Yang, Y.; Zhang, J.; Lan, C.; Li, Y.; Huang, L.; Huang, C.; Zhao, H.; Duan, Q. Crosslinking strategies and functionalization modification approaches for chitosan-based hydrogels in food preservation applications: A review. *Food Hydrocoll.* **2026**, *175*, 112475. [CrossRef]
14. Afrifa, M.A.O.; Awuah, S.G. Preclinical, clinical, and commercialization of chitosan in the biomedical space. *Carbohydr. Polym.* **2026**, *371*, 124456. [CrossRef] [PubMed]

https://doi.org/10.3390/polym18070812

polymers

MDPI

Article

Enhancing Antimicrobial and Antioxidant Properties of Chitosan-Based Films with 1-Methylimidazolium-Chitosan

Carolina Muñoz-Nuñez [1,2], Yoleida Quiroz-Pereira [1], Alexandra Muñoz-Bonilla [1,3,*] and Marta Fernández-García [1,3,*]

1 Instituto de Ciencia y Tecnología de Polímeros (ICTP-CSIC), C/Juan de la Cierva 3, 28006 Madrid, Spain
2 Facultad de Ciencias Químicas, Universidad Complutense de Madrid, Avenida Complutense s/n, Ciudad Universitaria, 28040 Madrid, Spain
3 Interdisciplinary Platform for Sustainable Plastics Towards a Circular Economy-Spanish National Research Council (SusPlast-CSIC), 28006 Madrid, Spain
* Correspondence: sbonilla@ictp.csic.es (A.M.-B.); martafg@ictp.csic.es (M.F.-G.)

Abstract

The design and the synthesis of functional films with enhanced functionality represent a significant step forward in sustainable material development due to their potential applications. In this study, a novel chitosan derivative (CS-MeIm) was synthetized by chemically modifying chitosan (CS) structure with 1-methyl-1H-imidazole (MeIm), a heterocyclic compound known for its biological properties. This functionalization not only enhances the intrinsic capabilities of CS but also provides a strategic platform for advanced material engineering. The modified compound, CS-MeIm, was incorporated at 10 wt% into films based on CS matrix, which was also reinforced with 1 or 5 wt% of chitin nanowhiskers (ChNw), to improve their functionality for its potential applications. The fabrication process was optimized to ensure the homogeneity and the structural integrity of the films, which were extensively evaluated to study their thermal stability, mechanical integrity, and bioactivity. The incorporation of the imidazole ring into the CS backbone provided a marked enhancement in antioxidant capacity from 3 to 15 μmol Trolox/gram of film; and excellent antimicrobial activity against common microbes, particularly against E. coli with an efficacy of 99.999%. The findings reveal that this chemical modification not only raises the intrinsic properties of CS but also introduces a versatile platform for creating biodegradable films with high functionality.

Keywords: chitosan; imidazole; chitin nanowhiskers; chemical modification; antimicrobial; antioxidant; films

1. Introduction

The proliferation of pathogenic microorganisms presents a significant challenge in sectors such as public health [1] and the food industry [2]. The increasing resistance to antibiotics underscores the urgent need for innovative, effective, and sustainable antimicrobial strategies [3,4]. In this context, materials with advanced functional properties emerge as promising solutions to fight microbes while ensuring environmental safety.

Heterocyclic compounds, such as pyridines [5], pyrroles [6], or imidazoles [7], have demonstrated great potential due to their unique structures and bioactive characteristics [8,9]. Specifically, the imidazole molecule, whose ring contains two nitrogen atoms and an aromatic system, exhibits antioxidant and antimicrobial properties that make it a valuable component in some formulations [10]. The imidazole ring can contribute to

 https://doi.org/10.3390/polym17192608

antioxidant activity through electron donation, thereby preventing oxidative damage, a crucial advantage in oxidation-sensitive applications like food packaging materials [11]. In addition, its structure enables electrostatic interactions and bonding with biomolecules, which is relevant for enhancing biocompatibility and functionality. When one of the nitrogen atoms in the imidazole ring is quaternized [12], the resulting positive charge could reinforce the electrostatic interactions and produces a synergistic effect that enhances antioxidant activity. In terms of antioxidant activity, the quaternized system can interact with free radicals through electronic resonance and the positive charge of the imidazolium cation, stabilizing radical species and contributing to the integrity of materials exposed to oxidative conditions [13]. Simultaneously, this positive charge enhances its antimicrobial properties by facilitating interactions with bacterial membranes, which have negatively charged surfaces. This electrostatic interaction disrupts the microbial membrane, allowing the entry of active compounds and ultimately inhibiting the growth of pathogenic bacteria [14].

The use of these molecules for polymer functionalization offers an effective way to enhance the biological properties of these materials [15–18]. The unique properties of imidazole and its derivatives could allow the creation of polymer systems with specific functions designed to meet the requirements of food active packaging applications, aiming to extend the shelf life of food products. Indeed, the demand for sustainable and active materials is growing, and this approach provides a pathway to address critical challenges related to food preservation and safety. In this context, Chitosan (CS) and derivatives have emerged as very promising biopolymers in food packaging area. CS, a biopolymer derived from the deacetylation chitin, is renowned for its biocompatibility [19] and biodegradability [20], and it is found in the exoskeletons of some crustaceans, crabs, or insects from natural sources, making it a sustainable and cost-effective option [21]. One of the most important characteristics of CS is its film-forming ability [22], enabling the formation of continuous and robust films that are ideal for packaging applications [23]. Additionally, its structure contains amino groups that can be easily chemically modified, allowing for versatile and adaptable functionalization [24,25], improving its bioactive properties, including enhanced oxidation resistance and antimicrobial capacity [18], two critical aspects for prolonging the shelf life of food products. Mechanical and barrier properties are also key aspects in food packaging; films and some nano-scale reinforcements have emerged as valuable components in polymer matrices. Among these, chitin nanowhiskers (ChNw) produced by the acid hydrolysis of chitin powder represent an interesting option due to their dimensions such as needle-like morphology [26], large surface area [27], and strong hydrogen bonding capacity [28], which enables them to form robust interactions with CS matrix. The addition of ChNw enhances the mechanical stability and flexibility of the composite but also improves its barrier properties against oxygen and moisture [29], factors crucial for applications in packaging and preservation.

In this study, chitosan (CS) has been modified with 1-methyl-1H-imidazole (MeIm) to create a functional derivative to be used as an additive in the preparation of CS-based films, introducing some biological properties. Moreover, the influence of reinforcing ChNw on the structure and performance of the films will be evaluated, enhancing their stability and mechanical strength.

This combination of additives and reinforcements would provide a robust platform for developing innovative materials solutions to address the challenges of food safety and environmental sustainability, particularly in the field of active and biodegradable food packaging.

2. Materials and Methods

2.1. Materials

Chitosan from shrimp shells (deacetylation degree > 75%), chitin form shrimp shells (Ch), 1-methyl-1H-imidazole (MeIm, 99%), 2-bromoacetic acid (97%), glycerol (Gly, 99%), ethyl-3-(3-dimethylaminopropyl) carbodiimide (EDC, ≥98%), *N*-hydroxysuccinimide (NHS, 98%), 2-(*N*-morpholino) ethane sulfonic acid (MES solution), sodium chloride (NaCl), and 2,2-diphenyl-1-picrylhydrazyl (DPPH) were purchased from Merck (Darmstadt, Germany). All the organic solvents were of analytical reagent grade, acetonitrile (CH_3CN), hydrochloric acid (HCl, 37%), tetrahydrofuran (THF), and methanol (MeOH) were provided by Scharlau (Barcelona, Spain). Deuterated dimethyl sulfoxide (DMSO-d6) was acquired from Merck. Cellulose dialysis membranes (CelluSep T1) were purchased from Membrane Filtration Products, Inc. (Seguin, TX, USA). Chitin nanowhiskers (ChNw) were synthesized according to our previous work [30].

For the antibacterial evaluation, saline solution (NaCl, suitable for cell culture, BioXtra, (Seneffe, Belgium) and phosphate-buffered saline (PBS) powder (pH 7.4) were sourced from Merck. BBL Mueller–Hinton broth microbial growth medium was obtained from Becton, Dickinson, and Company (Heidelberg, Germany), while the 96-well microplates and Columbia agar plates enriched with 5% sheep blood were acquired from Thermo Fisher Scientific (Dreieich, Germany). The following bacterial strains were utilized: *Escherichia coli* (*E. coli*, ATCC 25922), *Listeria innocua* (*L. innocua*, ATCC 33090), *Staphylococcus epidermidis* (*S. epidermidis*, ATCC 12228), *Staphylococcus aureus* (*S. aureus*, ATCC 29213), *Methicillin-resistant Staphylococcus aureus* (*MRSA*, ATCC 43300), *Enterococcus faecalis* (*E. faecalis* ATCC 29212), *Pseudomonas aeruginosa* (*P. aeruginosa*, ATCC 27853), and *Candida albicans* (*C. albicans*, ATCC 200955), all obtained from Oxoid™ (Thermo Fisher Scientific, Basingstoke, UK).

2.2. Synthesis of 3-(Carboxymethyl)-1-methyl-1H-imidazol-3-ium Bromide (MeImB)

First, 50 mmol of MeIm were dissolved in 25 mL of acetonitrile. Once dissolved, 50 mmol of bromoacetic acid were added to obtain the quaternized ammonium with a bromide ion. The reaction was stirred under an inert argon atmosphere and maintained at reflux for 24 h [31]. After this period, the solvent was removed using a rotary evaporator. The white solid was redissolved in the minimum possible amount of methanol, precipitated into an excess of cold tetrahydrofuran, and recovered by filtration. The MeImB product (see Scheme 1) was characterized by ^{1}H-NMR and ^{13}C-NMR.

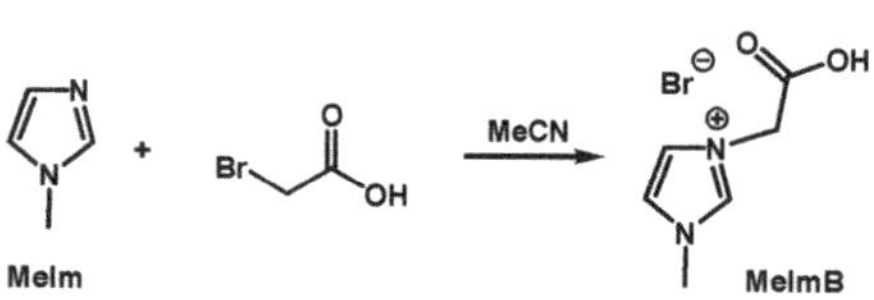

Scheme 1. Synthesis of 3-(carboxymethyl)-1-methyl-1H-imidazol-3-ium bromide (MeImB).

2.3. Synthesis of Chitosan Derivative (CS-MeImB)

Using the recently synthesized compound, MeImB was incorporated into CS through N-acylation, employing the EDC/NHS coupling system (Scheme 2). First, 0.5 g (1 meq of amine groups) of CS were dissolved in a 1% (v/v) aqueous glacial acetic acid solution. Simultaneously, 1.35 g of MeImB (2 meq) were dissolved in 15 mL of MES buffer solution (0.1 M) along with 1.18 g of EDC (2 meq) and stirred for 20 min [32]. After this period, 0.7 g of NHS (2 meq) were added, and stirring continued for 10 additional min. After this time, the MeImB solution was dropwise added to the CS solution, and the reaction was stirred for 24 h at room temperature [33]. After the corresponding dialysis process against water for 3 days (MW_{CO} of 12 kDa) and subsequent lyophilization, CS-MeImB was obtained and characterized with several techniques.

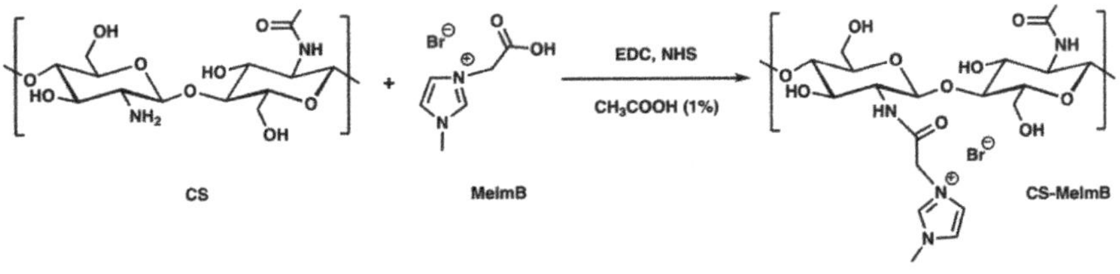

Scheme 2. Synthesis of chitosan derivative (CS-MeImB).

2.4. *Synthesis and Characterization of Chitin Nanowhiskers (ChNw)*

Chitin powder was acidified via hydrolysis in aqueous media to obtain ChNw. Briefly, 10 g of Ch were suspended in 500 mL of 3 M hydrochloric acid at 100 °C for 90 min under reflux conditions with continuous mechanical agitation. The acid suspension was then diluted in deionized water and subjected to three sequential centrifugation cycles of 10 min (12,000 rpm) to eliminate residual acid. The purified nanowhiskers were dialyzed (6–8 kDa MW_{CO} membrane) against distilled water for 4 days to neutralize pH, followed by freeze-drying to yield the final amount of ChNw [30]. The morphology of ChNw was analyzed using Field Emission Scanning Electron Microscopy (FE-SEM) on a Hitachi SU 8000 (Hitachi High-Technologies, Tokyo, Japan) in transmitted electron imaging mode. ChNw was dispersed in distilled water, and a drop was deposited onto a carbon-coated copper grid and allowed to dry at room temperature. The dried sample was subsequently coated with gold/palladium (80/20) before visualization. The longitudinal dimension of the nanowhiskers was measured using ImageJ 1.53K.

2.5. *Film Preparation*

Once the CS derivative was synthesized, it was used as an additive in the preparation of CS-based films, reinforced with previously obtained ChNws [30]. The films were produced with a base of 85% (*w*/*w*) polymer and 15% (*w*/*w*) glycerol as a plasticizer [34]. The polymer part was dissolved in a 1% (*v*/*v*) glacial acetic acid solution to obtain a final concentration of 1% (*w*/*v*). This solution included CS as the base matrix, the CS-MeImB additive, and varying amounts of ChNws to reinforce the film. All component quantities are detailed in Table 1. Glycerol was added to the solution, and the mixture was stirred for 24 h at room temperature. After this time, 50 mL of each solution were transferred to a Petri dish and allowed to dry by casting, resulting in the required films.

Table 1. Composition of chitosan-based films (%).

Sample	CS (%)	CS-MeImB (%)	ChNws (%)	Gly (%)
F-CS	85	-	0	15
F-CS-1	84	-	1	15
F-CS-5	80	-	5	15
F-CS-MeImB	75	10	0	15
F-CS-MeImB-1	74	10	1	15
F-CS-MeImB-5	70	10	5	15

2.6. *Characterization*

2.6.1. Nuclear Magnetic Resonance (NMR) Spectroscopy

NMR spectra (^1H and ^{13}C) were recorded at room temperature using a Bruker Avance III HD-400AVIII spectrometer (Bruker, Rheinstetten, Germany). The MeImB molecule was analyzed using DMSO-d6 as the solvent, while a 1% CD_3COOD solution in D_2O was used for the analysis of both polymers, CS and CS-MeImB.

2.6.2. Infrared Spectroscopy

The infrared spectra analysis was performed using a Perkin Elmer Spectrum Two instrument (PerkinElmer, Waltham, MA, USA) equipped with an attenuated total reflection module (ATR-FTIR). A background spectrum was collected before each sample measurement. Infrared spectra were recorded over a spectral range of 450 to 4000 cm^{-1} with a resolution of 4 cm^{-1}.

2.6.3. Elemental Analysis

A LECO CHNS-932 model PNT01 analyzer (LECO Corporation, St. Joseph, MI, USA) was utilized to perform the elemental analysis, which quantified the percentages of carbon, nitrogen, sulfur, and hydrogen in the samples. The acquired data, the mass ratios between carbon and nitrogen (C/N), were then used to determine the degrees of acetylation (DA) in chitosan and the degree of substitution (DS) in the chitosan modified with MeImB (CS-MeImB) (Equations (1) and (2)) [35].

$$\frac{C}{N} = \frac{(100 \times n_1 \times M_c) + (DA \times n_2 \times M_c)}{(100 \times n_3 \times M_N)} \quad (1)$$

$$\frac{C}{N} = \frac{(100 \times n_1 \times M_c) + (DA \times n_2 \times M_c) + (DS \times n_4 \times M_c)}{(100 \times n_3 \times M_N) + (DS \times n_5 \times M_N)} \quad (2)$$

where M_C and M_N are the molar masses of carbon and nitrogen, respectively; $n_1 = 6$, $n_2 = 2$ and $n_4 = 4$ are the number of carbons of deacetylated CS, acetamido group, and imidazole incorporated to Cs, respectively; and $n_3 = 1$ and $n_5 = 2$ are the number of nitrogen of CS and imidazole group, respectively.

2.6.4. Thermogravimetric Analysis

Thermogravimetric analysis was performed using a TGA Q500 analyzer (TA Instruments, New Castle, DE, USA). The thermograms were obtained from room temperature to 800 °C at a heating rate of 10 °C/min in an air atmosphere (50 cm^3/min). Calibration of the instrument for both temperature and weight was carried out following standard procedures.

2.6.5. Differential Scanning Calorimetry

The thermal behavior of the CS and its derivative, CS-MeImB, was studied with a TA Instruments (New Castle, DE, USA) DSC Q2000 differential scanning calorimeter, equipped with a refrigerated cooling system under N_2 atmosphere. The measurements were equilibrated at 0 °C and heated to 400 °C at 10 °C/min. The temperature scale was calibrated from the melting point of high-purity chemicals (lauric and stearic acids, and indium). Samples were prepared in sealed aluminum pans and weighed in an electronic balance.

2.6.6. Zeta Potential Measurements

The zeta potential of the CS and CS-MeImB was measured at physiological pH and 25 °C using a Zetasizer Nano series ZS (Malvern Instruments Ltd., Worcestershire, UK). For the preparation of the samples, the polymers were dissolved in the aqueous medium at a concentration of 1 mg/mL. The mixture was vortexed and sonicated to ensure homogeneity prior to analysis. The Smoluchowski equation was used to calculate electrophoretic mobility. The results are reported as the average from ten separate measurements.

2.6.7. X-Ray Diffraction

X-ray diffraction measurements were performed on CS and CS-MeImB at room temperature in reflection mode using a Bruker D8 Advance diffractometer equipped with a PSD Vantec detector (Bruker, Madison, WI, USA). CuKα radiation (λ = 0.15418 nm) was employed, operating at a voltage of 40 kV. The samples were locked on an appropriate holder and scanned over a range from 5° to 55° (2θ) with a step size of 0.02° and a collection time of 10 s per step.

2.6.8. Yellowness Index

The yellowness index (YI) of the films was evaluated following the ASTM D1925-70 standard [36]. The transmittance measurements were carried out using a PerkinElmer LAMBDA 1050+ UV/Vis/NIR spectrophotometer (PerkinElmer, Waltham, MA, USA), and the YI was calculated using the following equation:

$$YI = \frac{100 \times (C_X X - C_Z Z)}{Y} \quad (3)$$

where X, Y, and Z represent the CIE tristimulus values, while C_X (1.3013) and C_Z (1.1498) are coefficients determined by the illuminant (D65) and the observation angle (10°).

2.6.9. Water Vapor Transmission

The water vapor transmission (WVT) of the F-CS and F-CS-MeImB films was evaluated to determine their potential as moisture barriers, which is crucial for reducing microbial growth. Additionally, the influence of incorporating ChNw at concentrations of 1% and 5% was assessed.

WVT assay was conducted using ASTM E96-00e1 [37], the films studied were sealed over cups with a 10 mm diameter opening, containing 2 g of dry silica as a desiccant (relative humidity, RH 0%). These cups were then placed inside a desiccator maintained at 22 °C, containing a saturated sodium sulfate solution to establish a RH of 70%. The initial weight of each cup was recorded for 24 h, followed by periodic weight measurements to recorded changes in the moisture [38,39]. Three replicates of each film were analyzed to ensure statistical reliability. The weight against of time, and the WVT was determined from the slope of the linear regression, normalized by the exposed area and film thickness, using Equation (3). This normalization ensured an accurate comparison of permeability values across different film samples.

$$WVT = \left(\frac{\text{slope}(g/h)}{A(m^2) \times t(h)}\right) \times \text{thickness}(m) \quad (4)$$

2.6.10. Mechanical Properties

Tensile testing of the CS and CS-MeImB films was conducted utilizing a DX2000 QTest Elite MTS dynamometer (MTS Systems, Eden Prairie, MN, USA) at room temperature to evaluate their mechanical properties. Operating at a stretching rate of 10 mm/min and with a 100 N load cell. Dog-bone specimens with a gauge length of 20 mm and a width of 2 mm were used for the measurements. Stress–strain measurements were used to calculate the ultimate tensile strength, Young's modulus, and elongation at break values. Each film was tested with a minimum of five specimens.

2.6.11. Antioxidant Properties

The antioxidant activity of the CS and CS-MeImB was evaluated using the DPPH (2,2-diphenyl-1-picrylhydrazyl) method. A concentration of 1 mg of each sample was placed in 1 mL of DPPH solution. The DPPH solution was prepared by dissolving 6 mg of

DPPH in 250 mL of methanol, stored at 4 °C, and protected from light to maintain stability. The absorbance of the DPPH solution was measured at 517 nm using a BioTek® SYNERGY HTX multi-mode reader spectrophotometer (Agilent Technologies, Winooski, VT, USA) before and after the addition of the samples, at different times. The decrease in absorbance was used to calculate the percentage of DPPH scavenging activity, inhibition percentage, according to the following Equation (4):

$$I(\%) = \left(\frac{A_{control} - A_{sample}}{A_{control}}\right) \times 100 \tag{5}$$

where $A_{control}$ is the absorbance of the control (DPPH solution without samples), and A_{sample} is the absorbance of the DPPH solution after adding the samples.

In the same way, the antioxidant capacity of both CS and CS-MeImB films was studied using the same DPPH method. 1 mL of DPPH solution was added to a sample of 10 mg of each film. The resulting mixture was incubated for a specific period and was measured at different times at 517 nm to see the interaction between the DPPH radicals and the films.

To standardize the results, radical scavenging activities of the samples were expressed as equivalents of Trolox. A calibration curve was generated by plotting Trolox concentrations against their corresponding inhibition percentages (I%), allowing for the quantification of the antioxidant capacity of the tested materials.

2.6.12. Antimicrobial Properties

The antimicrobial properties of CS and CS-MeImB were assessed using the Minimum Inhibitory Concentration (MIC), following a standard broth dilution method according to the Clinical Laboratory Standards Institute (CLSI) [40]. The following microbial strains were used in the study: *S. aureus*, *S. epidermidis*, MRSA, *L. innocua*, *E. faecalis*, *E. coli*, *P. aeruginosa* and *C. albicans*. Bacterial and fungal strains were grown on 5% sheep blood Columbia agar plates at 37 °C for 24 h and 48 h, respectively.

Initially, microbial suspensions were freshly generated in saline solution to obtain a final concentration of 10^8 colony-forming units (CFU) per mL, corresponding to a turbidity near 0.5 on the McFarland scale. These initial suspensions were diluted to a concentration of ca. 10^6 CFU/mL using fresh Mueller-Hinton broth. For this purpose, 2 mg of each sample (CS and CS-MeImB) were dissolved in 1 mL of Mueller–Hinton broth. In a 96-well round-bottom microplate, 100 µL of this polymer solution was added into the first well, followed by the broth addition of 50 µL to the other wells. Serial dilutions were performed across the plate at a ratio of 1:2. Following this, 50 µL of the microbial suspension was introduced into each well, leading to a total volume of 100 µL and a final microbial concentration of ca. 5×10^5 CFU/mL.

The microplates were then incubated at 37 °C for 24 h and 48 h for bacterial and fungal strains. After incubation, the MIC was determined by visually inspecting the wells for bacterial growth. The lowest concentration of the sample that exhibited no visible growth was recorded as the MIC for each bacterial strain. All evaluations were conducted in at least three replicates.

Subsequently, the antimicrobial properties of the CS and CS-MeImB films were assessed using a similar methodology adapted from the E2149-20 standard method of the American Society for Testing and Materials (ASTM) [41] against the same microbial strains. 10 mg of each film were placed into 9 mL of PBS. Following this, 1 mL of a microbial suspension of ca. 10^6 CFU/mL was added to the film solution. The tubes were shaken at 100 rpm for 24 h. After incubation, the resulting solutions were grown on 5% sheep blood Columbia agar plates at 37 °C for 24 h and 48 h for bacterial and fungal strains, respectively.

The reduction percentage was determined by comparison with the CFU of the control by the plate counting methodology. The measurements were made at least in triplicate.

2.7. Statistical Analysis

All experiments were conducted in triplicate to ensure consistency. A one-way analysis of variance (ANOVA) was performed to evaluate significant differences in properties among the various groups, followed by post hoc Tukey's test, with a significance level set at $p \leq 0.05$.

3. Results and Discussion

Initially, the molecular weight of the CS was measured by viscometry according to the literature [42], the intrinsic viscosity of CS was measured in 0.1 M acetic acid with 0.2 M NaCl at different concentrations. The Mark–Houwink equation $[\eta] = KMw^a$ was used to calculate the molecular weight. The value of K and a of 1.81×10^{-3} cm^3/g and 0.93, respectively, were used [43]. The intrinsic viscosity was determined as 264 ± 7 cm^3/g; therefore, the Mw is the range of 357 ± 10 kDa. Then, this CS was modified with MeImB. The successful incorporation of MeImB was confirmed through spectroscopic techniques as will be shown below. This modification not only altered the chemical structure of the polymer but also resulted in a significant improvement in its physicochemical properties. The resulting polymer presents excellent solubility in aqueous media, achieving the primary objective of the modification. Specifically, the zeta potential of the modified compound was determined to be 49.2 ± 4.7 mV, indicating a highly positive surface charge.

3.1. Synthesis and Characterization of Chitosan Derivative

Firstly, the compound 3-(carboxymethyl)-1-methyl-1H-imidazol-3-ium was successfully synthesized, and its chemical structure was confirmed by ^{1}H NMR and ^{13}C NMR spectroscopies (see Figure 1). For the MeImB molecule, the ^{1}H NMR signals are observed at: δ (ppm) 13.7 (s, 1H, COOH), 9.1 (d, J = 1.6 Hz, 1H, H_4), 7.7 (q, J = 1.7 Hz, 2H, H_1 and H_2), 5.1 (s, 2H, H_5), and 3.9 (s, 3H, H_3) and the ^{13}C NMR signals are displayed at δ (ppm) 168.7 (C_6), 137.9 (C_4), 124.1 (C_1), 123.5 (C_2), 59.9 (C_5), 36.4 (C_3).

After characterizing MeImB compound, the chemical modification of chitosan was performed. ^{1}H NMR spectra of CS and its derivative were analyzed (see Figure 2). In the case of CS, a residual proton signal from the acetyl group is observed at 2.04 ppm overlapping with the methyl group of deuterated acetic acid, and a characteristic signal at 3.1 ppm corresponding to the H_2 position of the polymer ring. The H_1 signal overlaps with the D_2O solvent peak, while signals from H_3 to H_6 appear in the range of 4.0 to 3.3 ppm [44].

In the case of CS-MeImB, the polysaccharide structure exhibits similar signals to those of CS and additional signals assigned to the imidazole ring, mainly at 7.45 ppm and at 2.58 ppm that corresponds to the three protons of CH_3 [45,46]. These findings confirm the presence of new signals associated with the synthesized compound.

3.2. FTIR Analysis

The chemical modification of CS with methyl imidazole was also studied by FTIR, the spectra of native CS, MeImB and the resulting CS-MeImB derivative are shown in Figure 3. A broad band at 3340 cm^{-1} corresponding to the –NH and –OH stretching vibrations and the intramolecular hydrogen bonds of the chitosan, followed by a band at 2870 cm^{-1} attributed to the stretching vibration of –CH groups [47] were observed. Two characteristic bands at 1650 cm^{-1} and 1560 cm^{-1} were seen, corresponding to the absorption of the –NH–CO– stretching vibration and the deformation vibration of NH_2 groups, respectively [48].

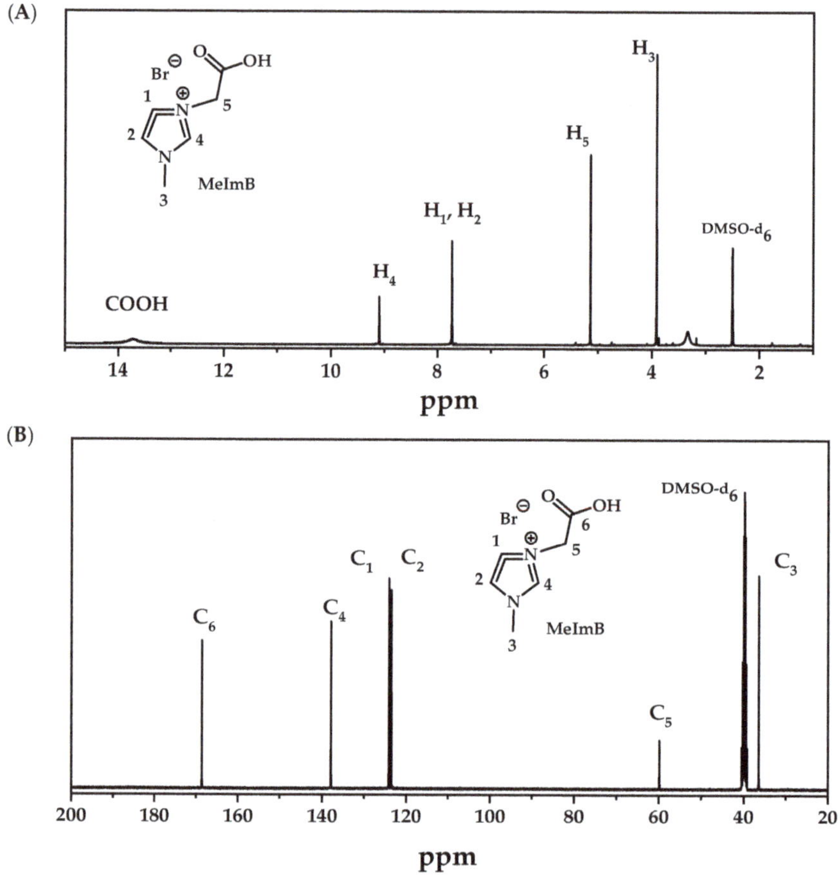

Figure 1. (**A**) ^{1}H NMR and (**B**) ^{13}C NMR spectra of MeImB.

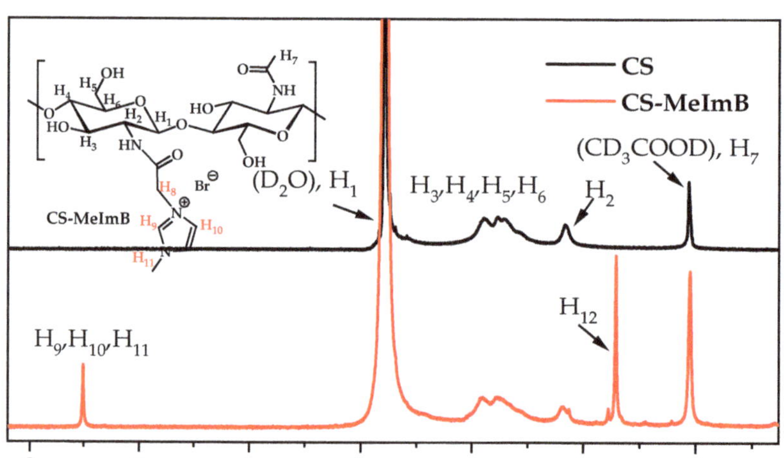

Figure 2. ^{1}H NMR spectra of CS and CS-MeImB.

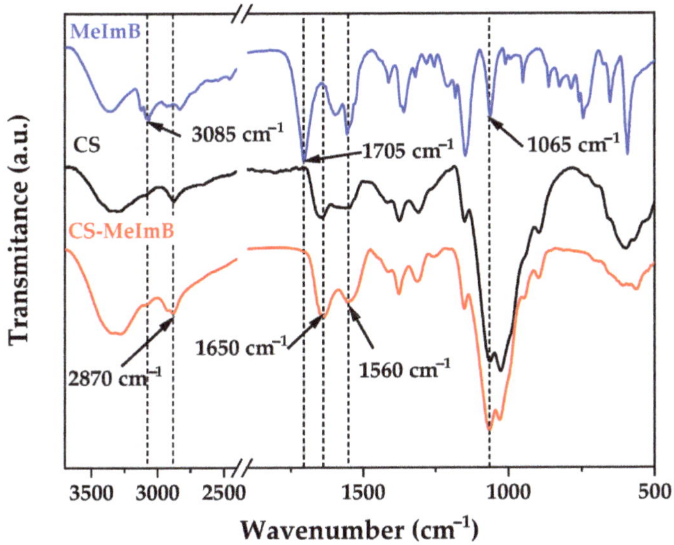

Figure 3. FTIR spectra of CS and CS-MeImB.

On the other hand, for CS-MeImB, the characteristic bands of the CS were still observed, along with additional bands corresponding to the modification. A new band appeared at 3085 cm^{-1}, corresponding to the stretching vibration of the –CH groups from the introduced methyl imidazole ring. Furthermore, the bands at 1650 cm^{-1} and 1560 cm^{-1}, associated with the inserted amide groups and the deformation vibration of –NH–, both showing increased intensity and exhibiting a slight shift, due to the amino group of the imidazole ring [49].

3.3. Elemental Analysis

The degree of modification was calculated based on elemental analysis, and the results are presented in Table 2. The theoretical values of CS elemental analysis are also included for clarity. MeImB molecule presents in its elemental analysis 51.1% for C, 6.4% for H, and 19.9% for N; therefore, a C/N ratio of 2.6. Then, a significant decrease in the C/N ratio was observed after the modification of chitosan attributed to the insertion of the imidazole group. From these data the degree of substitution was found to be 57%. It is worth noting that the degree of acetylation (DA) remains constant, as the CS was not subjected to a prior deacetylation process.

Table 2. Percentage of carbon, hydrogen, and nitrogen, degree of acetylation, and degree of substitution of CS and CS-MeImB.

	C (%)	H (%)	N (%)	C/N	DA (%)	DS (%)
CS	40.6 ± 0.3	6.7 ± 0.3	7.2 ± 0.3	5.6	28	-
CS_{theo}	44.3	6.9	8.0	5.5	25	
CS-MeImB	39.5 ± 0.3	6.5 ± 0.1	9.4 ± 0.1	4.2	28	57

3.4. X-Ray Diffraction (XRD)

The X-ray diffraction patterns of CS and CS-MeImB are presented in Figure 4. For CS, two prominent peaks were observed at 9.4° and 20.0°, corresponding to the (020) and (110) planes of the crystalline structure, respectively [50].

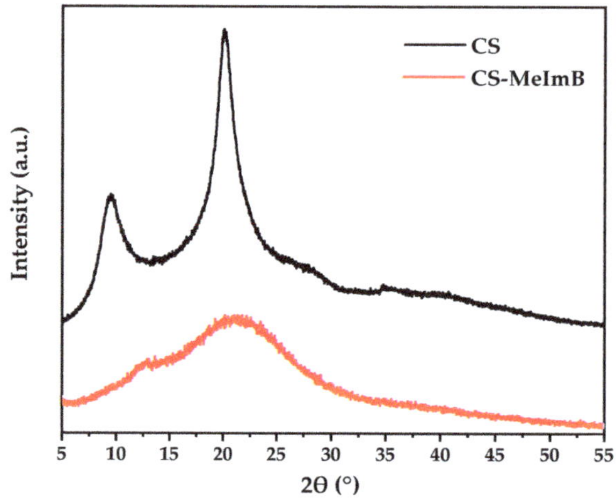

Figure 4. X-ray diffraction patterns of CS and CS-MeImB.

After modification, the crystalline phase is destroyed; only a very small peak at 14° was observed besides the wider amorphous halo [51]. This could suggest that the imidazole ring retains some crystallinity within the modified structure. However, introducing polar groups disrupts the strong hydrogen bond between CS chains, inducing the disorganization of crystalline structure [52]. Consequently, the material exhibits a higher amorphous content and a reduced crystalline phase. The introduction of the polar molecule promotes strong covalent and non-covalent interactions among the hydroxyl functional groups of CS [53].

3.5. Thermogravimetric Analysis (TGA)

The thermogravimetric analysis of CS and CS-MeImB revealed distinct thermal behavior for both samples (Figure 5). An initial weight loss of approximately 10% was observed in both cases, attributed to water loss, occurring between 40 °C and 150 °C, which is typical for highly hydrophilic polysaccharides [54].

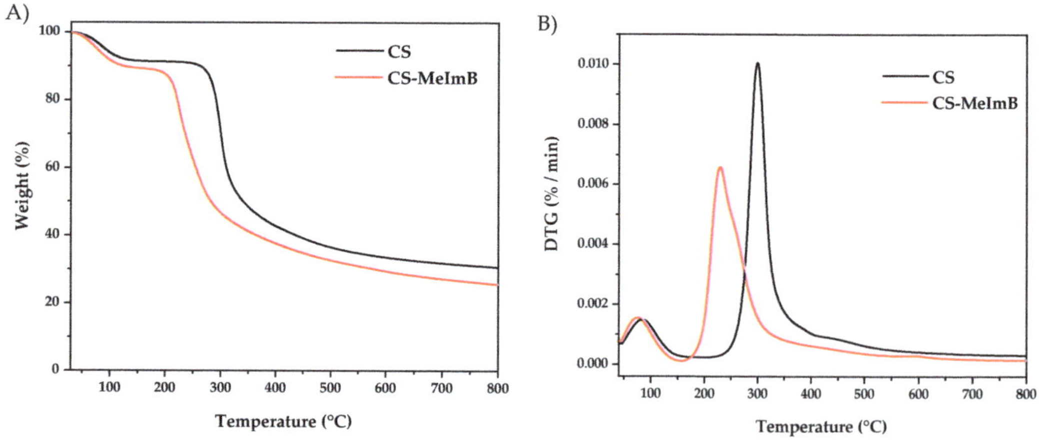

Figure 5. (**A**) Thermogravimetric curves of CS and CS-MeImB and (**B**) their derivatives (DTG).

Subsequently, in the case of the CS derivative, a second stage of thermal decomposition was evident, corresponding to the degradation of the organic matrix and the chemically attached methylimidazole groups, resulting in a 65% mass loss. This suggests that the introduced organic group as side chain with the formation of amide could be a starting site of breaking promotes the thermal degradation of CS, and decreasing the thermal stability of the final material [55,56]. For unmodified CS, the degradation stage of the glucosamine units occurred between 257 °C and 392 °C, leading to a total mass loss of approximately 60%. The carbonaceous residue is high due to the inert atmosphere used and is habitual found in polysaccharides and glycopolymers [57,58].

3.6. *Differential Scanning Calorimetry (DSC)*

The thermal properties of CS and CS-MeImB were analyzed using DSC to understand the effects of the modification (Figure 6). Both samples exhibit a broad endothermic peak between 48 °C and approximately 150 °C, attributed to the evaporation of water bound to the polymer. In the case of CS, an exothermic peak also appears between 280 °C and 340 °C, corresponding to the decomposition of the main backbone of the polymer [59].

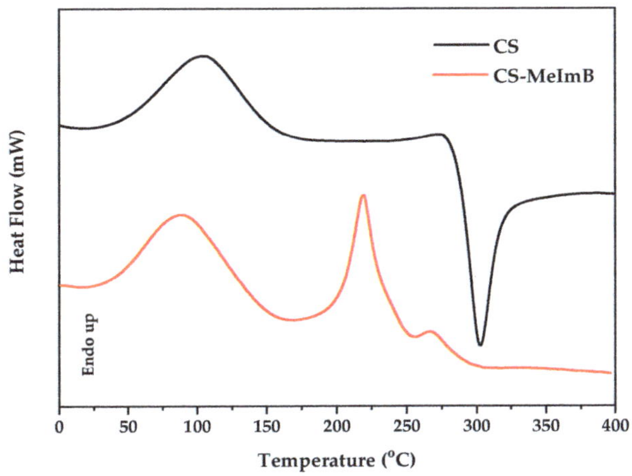

Figure 6. Differential scanning calorimetry curves of CS and CS-MeImB.

In the case of the CS-MeImB, the second peak appears around 214 °C as an endothermic process, corresponding to the decomposition of the complex formed between CS and the MeImB molecule [60,61]. Additionally, the formation of the derivative reduces the decomposition temperature of the CS backbone due to modifications in the natural organization of intermolecular bonds in CS-MeImB [62].

3.7. *Antioxidant Activity of Chitosan Derivative*

CS is known for its intrinsic antioxidant properties due to the presence of –OH– and –NH_2 groups, which can donate hydrogen atoms to neutralize reactive species [49,63,64]. However, chemical modifications could alter these properties by affecting hydrogen bonding and electron delocalization.

Imidazole molecules are known for their antioxidant properties due to their ability to stabilize free radicals through electron donation and resonance effects [10,65]. Additionally, the inherent positive charge of imidazolium-functionalized compounds could enhance antioxidant activity by facilitating interactions with radical species [9,18]. To evaluate this possible enhancement, the antioxidant activity was quantified using the DPPH assay, with Trolox as the standard for calibration.

Figure 7 illustrates the DPPH radical scavenging capacity of CS and CS-MeImB. It is clearly appreciated that the antioxidant capacity is enhanced in the derivative with the functionalization of MeImB. For CS, the radical stabilization reaches its maximum after

5 h, achieving 58–60 μmol of Trolox per gram of polymer [66]. In contrast, CS-MeImB demonstrates faster DPPH scavenging kinetics, reaching a plateau within the same time but with double the capacity, reaching 120 μmol of Trolox per gram of polymer.

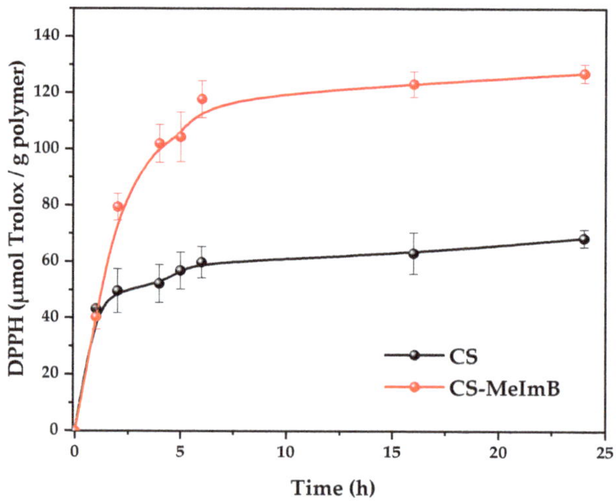

Figure 7. Antioxidant activity of CS and CS-MeImB versus time.

Heterocyclic groups have been extensively investigated to improve the antioxidant capacity of CS, with the imidazole group giving the most promising results, achieving nearly 90% inhibition at relatively low concentrations [67]. Thus, it can be concluded that the presence of the quaternary imidazole ring in CS-MeImB significantly improves the DPPH radical stabilization capacity, which explains its higher antioxidant activity and accelerated kinetics.

3.8. Antimicrobial Activity of CS-MeImB

After proper characterization of the chemical structure and properties of CS-MeImB, its potential antimicrobial activity was also evaluated. In contrast to native CS, which is insoluble in aqueous media and forms aggregates at physiological pH, CS-MeImB demonstrates significant solubility under these conditions. This improved solubility appears from its cationic structure and net positive charge, which enable better dispersion at physiological pH and provide antimicrobial capabilities [18,62]. Indeed, the zeta potential, a measure of surface charge density, was determined to be 49 ± 5 mV. As mentioned above, this high positive charge density enhances not only its water stability but also its ability to interact with the negatively charged components of bacterial cell walls, potentially leading to greater antimicrobial efficacy. The combination of physiological solubility and a pronounced positive charge suggests that CS-MeImB might exhibit superior antimicrobial properties compared to CS. Certainly, this derivative presents a strong effectiveness against Gram-positive and negative bacteria, including resistant strains, and also against fungi, as can be seen in Figure 8, with MIC values remarkably low for the majority of the tested microorganisms. Similarly, other chitin derivatives containing imidazolium groups reported in the literature exhibited broad-spectrum antimicrobial activity [24,49].

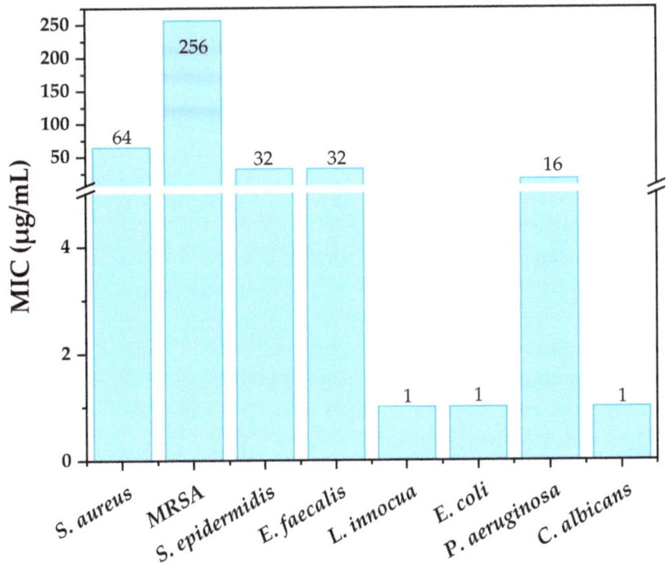

Figure 8. MIC values of CS-MeImB against different strains.

3.9. Characterization of Films

Once the new compound has been characterized, the next step involves the preparation of CS films incorporating CS-MeImB. This derivative was added as an active component, aiming to enhance the properties of the resulting films. Therefore, the films underwent a wide physicochemical and biological characterization.

3.10. Thermogravimetric Analysis of Films

The thermal decomposition of CS films containing CS-MeImB films, along with ChNws, was analyzed using TGA. The mass loss curves for all films and their derivative curves are shown in Figures 9A and 9B, respectively. Within the studied temperature range, two main degradation steps can be observed, along with a shoulder in the second step [68].

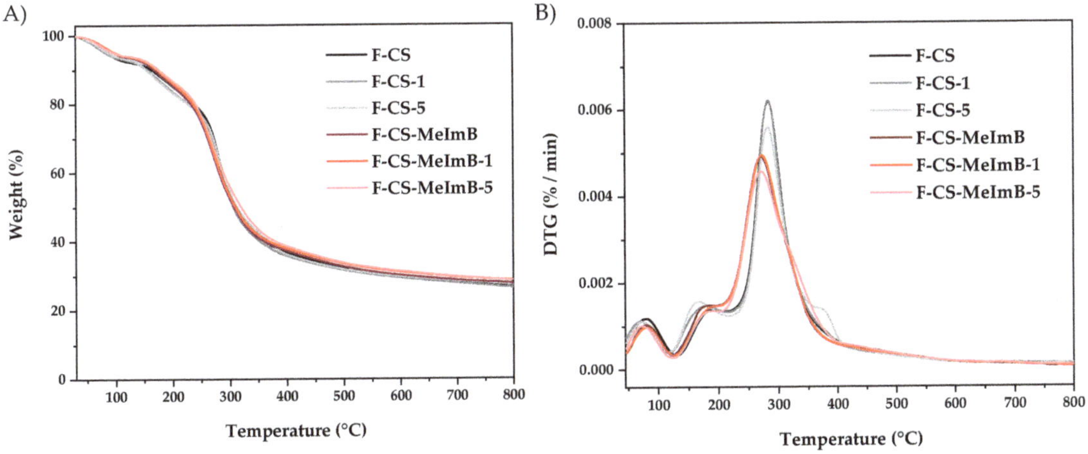

Figure 9. (**A**) Thermogravimetric curves of CS films and (**B**) its derivative (DTG).

In all cases, the first thermal event occurs from room temperature up to 120 °C, corresponding to the evaporation of absorbed water from the films due to their hydrophilic nature [69]. As the temperature increases, a shoulder appears around 190 °C, associated with glycerol evaporation, present in the formulations in a 15% [70]. This event is similar for all samples; however, in the case of CS-MeImB films, the shoulder is slightly shifted to a higher temperature. This shift could be attributed to the inherent positive charge of the MeImB moiety, which likely forms stronger hydrogen bonds with glycerol, requiring a higher temperature for its evaporation.

Finally, the second degradation step occurs between 220 °C and 350 °C, corresponding to the decomposition of CS main chain. Differences can be observed between the films made solely of CS and those containing 10% CS-MeImB, with the latter exhibiting lower thermal stability, as previously observed. Regarding the influence of ChNws incorporation, the addition of these nanocrystals did not seem to affect the thermal stability, presenting values of the maximum degradation temperature similar to that without ChNws. This fact was analogous in both CS and CS-MeImB films [71].

3.11. X-Ray Diffraction of the Films

X-ray diffraction analysis was performed to evaluate the crystalline structure of the films produced with CS and CS-MeImB. To further investigate the structural modifications induced by the chemical functionalization, the effect of incorporating the two different amounts of ChNws into the films was also studied. Films containing two different ChNw concentrations (1% and 5%) were analyzed to assess their influence on the crystallinity and potential interactions with the polymer matrix (Figure 10).

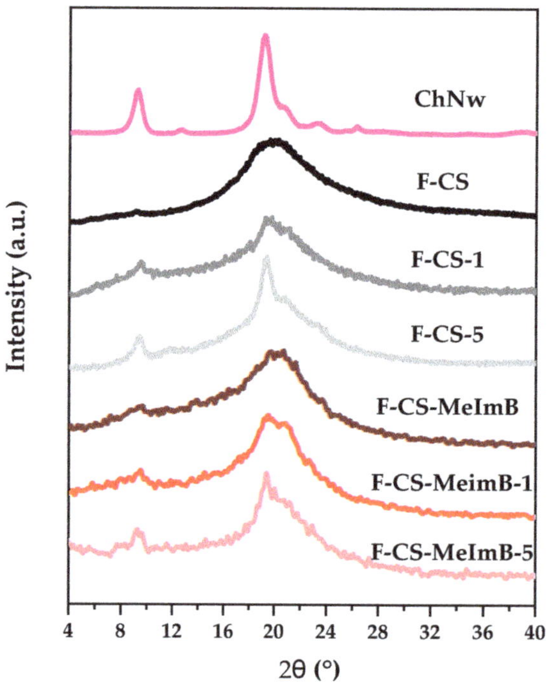

Figure 10. X-ray diffraction patterns of CS films.

First, the crystalline structure of ChNw was characterized by its distinct peaks at 2θ = 8.9° (020 plane) and 19° (110 plane), along with a shoulder at 2θ = 20° (120 plane).

This confirms that the internal crystalline structure of the nanowhiskers matches that of chitin powder [72].

Regarding the crystalline structure of the F-CS film, it appears predominantly amorphous, with a broad diffuse peak at $2\theta = 19.7°$ and a barely perceptible peak at $2\theta = 9.4°$ [73]. This suggests that dissolving CS in an acidic medium with Gly disrupts the formation of intramolecular hydrogen bonds, restricting the mobility of CS chains during film formation and consequently disturbing the crystalline regions of the polymer. When different amounts of ChNws were incorporated into pure CS films, the characteristic crystalline peaks of the nanocrystals at $2\theta = 8.9°$ and 19° became more pronounced as their concentration increased, while the amorphous structure of the F-CS film with the plasticizer remained unchanged [74].

Incorporating CS-MeImB into the polymer matrix does not introduce notable structural changes. First, the peak at $2\theta = 9.4°$ is maintained in both of the cases, with and without CS-MeImB. Also, the broad band at $2\theta = 19.7°$ remains present, confirming the amorphous nature of the material [75]. Furthermore, as previously observed, the addition of ChNws enhances the intensity of its characteristic crystalline peaks, resulting in a more crystalline overall film [76].

3.12. Yellowness Index

The yellowing index (YI) was analyzed in the films due to its importance in preserving visual quality, ensuring product integrity against external factors, and meeting industry standards. UV-visible spectroscopy was employed to assess YI in the 380–780 nm range, using a 10 nm step width. Due to the inherent transparency of the films, transmittance measurements were used to evaluate changes in optical properties (see Table 3). The results indicate that the incorporation of ChNws led to an increase in film turbidity and a higher yellowing index, confirming that nanowhisker addition in a high proportion (5%) impacts the optical clarity of the films. However, when comparing CS and CS-MeImB films, no significant differences in YI were observed. This suggests that the chemical modification did not substantially alter the optical properties of the films.

Table 3. Yellow indexes (YI), water vapor transmission (WVT), mechanical properties of CS films elastic modulus (E), the maximum tensile strength (TS), and elongation at break (ε).

	YI (%)	WVT 10^{-3} (g/(d × m)	E (MPa)	TS (MPa)	ε (%)
F-CS	2.73 ± 0.16^{a}	81.4 ± 5.4^{a}	11.2 ± 1.6^{a}	28.7 ± 3.5^{a}	10.3 ± 3.0^{a}
F-CS-1	3.16 ± 0.15^{a}	78.6 ± 5.5^{a}	13.9 ± 2.1^{a}	24.0 ± 4.9^{a}	6.4 ± 1.5^{ab}
F-CS-5	6.34 ± 0.19^{b}	70.4 ± 8.7^{a}	19.0 ± 2.1^{b}	24.3 ± 5.7^{a}	8.9 ± 3.8^{ab}
F-CS-MeImB	2.78 ± 0.15^{a}	21.1 ± 0.9^{b}	14.1 ± 0.8^{a}	22.4 ± 5.3^{a}	3.6 ± 2.1^{c}
F-CS-MeImB-1	3.13 ± 0.16^{a}	22.6 ± 1.5^{b}	15.7 ± 1.0^{ab}	28.0 ± 2.1^{a}	2.8 ± 1.2^{c}
F-CS-MeImB-5	5.09 ± 0.94^{b}	23.5 ± 0.9^{b}	19.5 ± 1.8^{b}	26.8 ± 6.1^{a}	1.8 ± 0.6^{c}

Values with the same letter are not significantly different for Tukey test (Significance level of $p \leq 0.05$).

From an industrial perspective, maintaining a low yellowing index is advantageous in packaging applications, as excessive yellowing may be undesirable to consumers. The results indicate that, despite the increased turbidity caused by ChNws, the overall visual appearance of the films remains within an acceptable range, making them suitable for potential packaging applications.

3.13. Water Vapor Transmission

The WVT of CS films and those incorporating CS-MeImB were evaluated with the different amounts of ChNws. In all cases, the WVT values were higher for CS films compared to films containing CS-MeImB (see Table 3). This difference could be attributed to the strong and active intermolecular interactions in the case of CS-MeImB derivative, which restrict the diffusion of water vapor through the polymeric matrix. The reduced WVT could also be related to the presence of the positively charged imidazole group, which promotes additional interactions, leading to a more compact and less permeable structure [77]. Furthermore, in both types of films, the WVT values do not vary with these ChNws content.

3.14. Mechanical Properties

The mechanical properties of the films were evaluated in terms of Young's modulus, tensile strength, and elongation at break (see Table 3). The effect of incorporating 1% and 5% ChNws into both F-CS and F-CS-MeImB films was analyzed to determine their impact on film stiffness, strength, and flexibility.

For native CS based films, the addition of nanowhiskers led to an increase in Young's modulus from 11.2 ± 1.6 MPa (F-CS) to 19.0 ± 2.1 MPa (F-CS-5), after the addition of ChNw, indicating enhanced stiffness due to the reinforcement effect. However, no significant changes were observed in tensile strength, remaining in the range of 24–29 MPa and suggesting that the intermolecular interactions between CS and ChNws did not considerably alter the film's maximum resistance to deformation. Similarly, elongation at break remained unchanged, implying that the overall flexibility of the films was not significantly affected by the incorporation of nanowhiskers, as reported in the literature [78].

A similar trend was observed for the containing CS-MeImB films, where the incorporation of nanowhiskers at both 1% and 5% increased Young's modulus, from 14.1 ± 0.8 MPa (F-CS-MeImB) to 19.5 ± 1.8 MPa (F-CS-MeImB-5); while tensile strength and elongation at break remained largely unaffected between 22 and 27 MPa. However, a comparison between CS and CS-MeImB films revealed some small differences. The Young's modulus values were a bit higher for the F-CS-MeImB, suggesting that the chemical modification contributed to the increase in stiffness. In contrast, tensile strength values remained comparable between both film types, while elongation at break was considerably lower in CS-MeImB films compared to unmodified CS films.

The observed increase in stiffness with the incorporation of ChNws is aligned with previous studies, which have attributed this effect to hydrogen bonding interactions between polymer chains and the reinforcing nanowhiskers [79]. This reinforcement mechanism limits molecular mobility, resulting in stiffer films. Moreover, the lack of significant changes in tensile strength suggests that the interfacial adhesion between chitosan and chitin nanowhiskers is sufficient to maintain the film's mechanical integrity without introducing substantial structural defects. The observed decrease in elongation at break in CS-MeImB films may be attributed to the combined effects of chemical modification and nanoparticles incorporation, which could lead to a more rigid polymer network with reduced flexibility.

3.15. Antioxidant Activity of Chitosan Films

The antioxidant activity of the films was evaluated to determine the effect of CS-MeImB and the incorporation ChNws on their free radical scavenging capacity. The influence of CS-MeImB on the CS matrix is expected to affect the antioxidant behavior of the films by enhancing interactions with free radicals. To assess these effects, the antioxidant activity was measured using the DPPH radical scavenging assay, comparing the different film compositions (Figure 11).

The radical scavenging capacity of films containing only CS was relatively low, approximately 3 μmol Trolox per gram of film, reaching equilibrium after 5 h in all cases (F-CS, F-CS-1 and F-CS-5), regardless of the amount of ChNws added, as the values remain statistically similar.

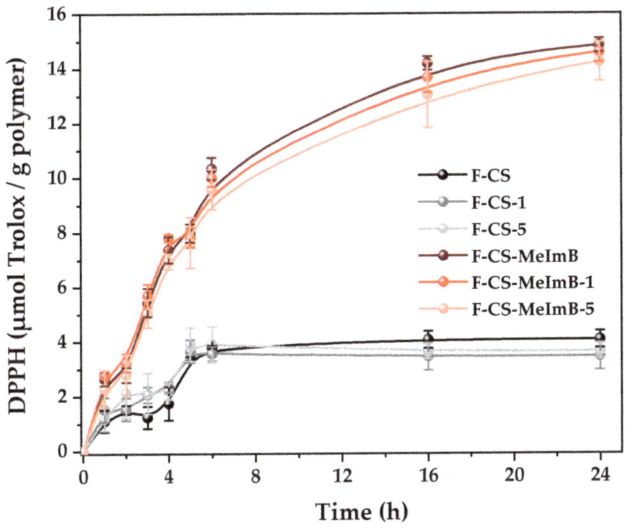

Figure 11. Antioxidant activity of CS films.

In contrast, films made with CS-MeImB demonstrated a significantly higher capacity for scavenging DPPH radicals, reaching 14–15 μmol Trolox per gram of film, after approximately 20 h. This enhanced antioxidant activity was attributed to the increased positive charge and the electronic delocalization of the imidazole ring [80]. However, the incorporation of 1% or 5% nanowhiskers in these films slightly reduces their antioxidant activity, possibly due to steric effects limiting the accessibility to the active sites.

3.16. Antimicrobial Activity of Chitosan Films

The evaluation of the antimicrobial activity of the films revealed clear differences between the unmodified CS films and those functionalized with MeImB (see Table 4). Films composed exclusively of CS exhibited insignificant antibacterial activity, except against *S. aureus*, where a moderate inhibitory effect was observed. However, no significant reduction in bacterial growth was detected for *P. aeruginosa*, *E. faecalis* or *E. coli*. In contrast, films containing active CS-MeImB demonstrated a great improvement in antimicrobial performance. This effect was particularly pronounced against Gram-negative bacteria, where complete bacterial elimination was observed in the case of *E. coli*. These findings highlight the effectiveness of MeImB-modified CS films, enhancing antimicrobial properties, particularly against Gram-negative bacterial strains. Examples of bacterial reduction in films are shown in Figure 12.

Table 4. Antibacterial activity of the resulting CS based films estimated as bacterial reduction percentage.

	S. aureus	*S. epidermidis*	*P. aeruginosa*	*E. coli*
F-CS	85.3 ± 1.1 [a]	-	-	-
F-CS-1	88.5 ± 1.4 [b]	-	-	-
F-CS-5	79.7 ± 0.9 [c]	-	-	-
F-CS-MeImB	85.5 ± 2.3 [ab]	99.5 ± 2.1 [a]	97.5 ± 1.7 [a]	99.999 ± 0.001 [a]
F-CS-MeImB-1	87.7 ± 2.7 [ab]	99.8 ± 1.3 [a]	99.1 ± 2.1 [a]	99.999 ± 0.001 [a]
F-CS-MeImB-5	92.1 ± 4.3 [ab]	97.8 ± 2.6 [a]	99.4 ± 1.5 [a]	99.999 ± 0.001 [a]

Values with the same letter are not significantly different for Tukey test (Significance level of $p \leq 0.05$).

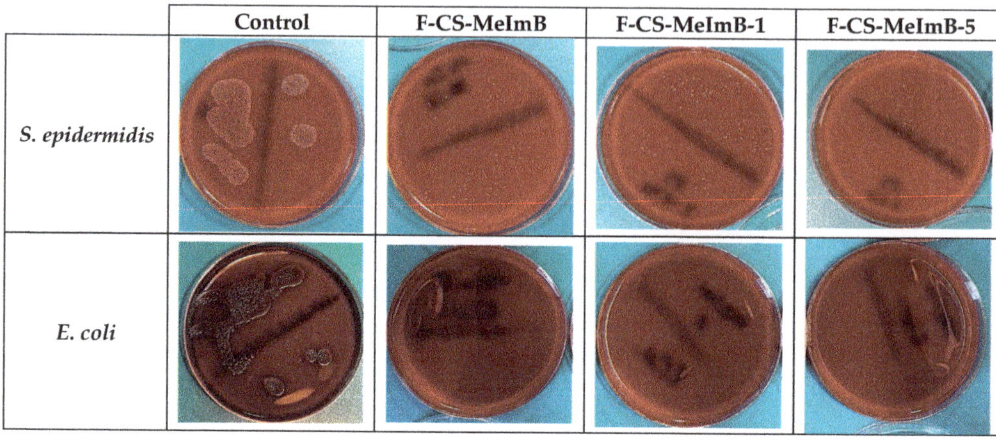

Figure 12. Images of the agar plates used for microbial colonies determination by plate counting method.

4. Conclusions

In this study, a water-soluble, strong antioxidant and antimicrobial imidazolium chitosan derivative, CS-MeImB, was obtained and incorporated as an active additive into CS-based films. The films were also reinforced with ChNws to improve mechanical and barrier properties. The functional CS films containing CS-MeImB demonstrated enhanced antioxidant properties and strong inhibitory effect against different strains, particularly against Gram-negative bacteria such as *E. coli* and *P. aeruginosa*. The incorporation of the CS-MeImB derivative slightly modified the mechanical properties, increasing Young's modulus but decreases elongation at break. In addition, they exhibited notably lower water vapor transmission values compared to CS films, indicating an enhanced barrier effect resulting from the chemical modification. All these improvements were obtained without diminishing their optical properties. On the other hand, the incorporation of ChNw as fillers did not significantly modified their mechanical performance but facilitated their barrier properties.

All these characteristics make them promising materials for biodegradable packaging and other advanced applications where structural reinforcement and bioactivity are required.

Author Contributions: Conceptualization: A.M.-B. and M.F.-G.; investigation: C.M.-N. and Y.Q.-P.; writing—original draft preparation: C.M.-N., A.M.-B. and M.F.-G.; writing—review and editing: C.M.-N., A.M.-B. and M.F.-G.; project administration and funding acquisition: C.M.-N., A.M.-B. and M.F.-G. All authors have read and agreed to the published version of the manuscript.

Funding: This work was funded by the MICIU (PID2022-136510B-100), the Agencia Estatal de Investigación (AEI, Spain) and Fondo Europeo de Desarrollo Regional (FEDER, EU). C.M.-N. also acknowledges MICINN for her FPI fellowship PRE2020-093596.

Institutional Review Board Statement: No applicable.

Data Availability Statement: The raw data supporting the conclusions of this article will be made available by the authors on request.

Conflicts of Interest: The authors declare no conflicts of interest.

References

1. Serwecińska, L. Antimicrobials and Antibiotic-Resistant Bacteria: A Risk to the Environment and to Public Health. *Water* **2020**, *12*, 3313. [CrossRef]

2. Siroli, L.; Patrignani, F.; Serrazanetti, D.I.; Chiavari, C.; Benevelli, M.; Grazia, L.; Lanciotti, R. Survival of Spoilage and Pathogenic Microorganisms on Cardboard and Plastic Packaging Materials. *Front. Microbiol.* **2017**, *8*, 2206. [CrossRef] [PubMed]
3. Brooks, B.D.; Brooks, A.E. Therapeutic Strategies to Combat Antibiotic Resistance. *Adv. Drug Deliv. Rev.* **2014**, *78*, 14–27. [CrossRef]
4. Namivandi-Zangeneh, R.; Wong, E.H.H.; Boyer, C. Synthetic Antimicrobial Polymers in Combination Therapy: Tackling Antibiotic Resistance. *ACS Infect. Dis.* **2021**, *7*, 215–253. [CrossRef]
5. Marinescu, M.; Popa, C.-V. Pyridine Compounds with Antimicrobial and Antiviral Activities. *Int. J. Mol. Sci.* **2022**, *23*, 5659. [CrossRef]
6. Abd El-Hameed, R.H.; Sayed, A.I.; Mahmoud Ali, S.; Mosa, M.A.; Khoder, Z.M.; Fatahala, S.S. Synthesis of Novel Pyrroles and Fused Pyrroles as Antifungal and Antibacterial Agents. *J. Enzym. Inhib. Med. Chem.* **2021**, *36*, 2183–2198. [CrossRef]
7. Sharma, D.; Narasimhan, B.; Kumar, P.; Judge, V.; Narang, R.; De Clercq, E.; Balzarini, J. Synthesis, Antimicrobial and Antiviral Evaluation of Substituted Imidazole Derivatives. *Eur. J. Med. Chem.* **2009**, *44*, 2347–2353. [CrossRef]
8. Rusu, A.; Moga, I.-M.; Uncu, L.; Hancu, G. The Role of Five-Membered Heterocycles in the Molecular Structure of Antibacterial Drugs Used in Therapy. *Pharmaceutics* **2023**, *15*, 2554. [CrossRef]
9. Tan, W.; Zhang, J.; Zhao, X.; Dong, F.; Li, Q.; Guo, Z. Synthesis and Antioxidant Action of Chitosan Derivatives with Amino-Containing Groups via Azide-Alkyne Click Reaction and N-Methylation. *Carbohydr. Polym.* **2018**, *199*, 583–592. [CrossRef]
10. Siwach, A.; Verma, P.K. Synthesis and Therapeutic Potential of Imidazole Containing Compounds. *BMC Chem.* **2021**, *15*, 12. [CrossRef] [PubMed]
11. Alizadeh Sani, M.; Khezerlou, A.; McClements, D.J. Zeolitic Imidazolate Frameworks (ZIFs): Advanced Nanostructured Materials to Enhance the Functional Performance of Food Packaging Materials. *Adv. Colloid Interface Sci.* **2024**, *327*, 103153. [CrossRef]
12. Van de Vliet, L.; Vackier, T.; Thevissen, K.; Decoster, D.; Steenackers, H.P. Imidazoles and Quaternary Ammonium Compounds as Effective Therapies against (Multidrug-Resistant) Bacterial Wound Infections. *Antibiotics* **2024**, *13*, 949. [CrossRef]
13. Zhao, L.; Zhang, C.; Zhuo, L.; Zhang, Y.; Ying, J.Y. Imidazolium Salts: A Mild Reducing and Antioxidative Reagent. *J. Am. Chem. Soc.* **2008**, *130*, 12586–12587. [CrossRef] [PubMed]
14. Riduan, S.N.N.; Zhang, Y. Imidazolium Salts and Their Polymeric Materials for Biological Applications. *Chem. Soc. Rev.* **2013**, *42*, 9055–9070. [CrossRef]
15. Romano, S.; De Santis, S.; Martinelli, A.; Rocchi, L.A.; Rocco, D.; Sotgiu, G.; Orsini, M. Starch Films Plasticized by Imidazolium-Based Ionic Liquids: Effect of Mono- and Dicationic Structures and Different Anions. *ACS Appl. Polym. Mater.* **2023**, *5*, 8859–8868. [CrossRef]
16. Anderson, E.B.; Long, T.E. Imidazole- and Imidazolium-Containing Polymers for Biology and Material Science Applications. *Polymer* **2010**, *51*, 2447–2454. [CrossRef]
17. Alqahtani, N.F. Functionalized Imidazolium Ionic Liquids-Modified Chitosan Materials: From Synthesis Approaches to Applications. *React. Funct. Polym.* **2024**, *194*, 105779. [CrossRef]
18. Wei, L.; Li, Q.; Chen, Y.; Zhang, J.; Mi, Y.; Dong, F.; Lei, C.; Guo, Z. Enhanced Antioxidant and Antifungal Activity of Chitosan Derivatives Bearing 6-O-Imidazole-Based Quaternary Ammonium Salts. *Carbohydr. Polym.* **2019**, *206*, 493–503. [CrossRef] [PubMed]
19. Hsu, S.; Chang, Y.-B.; Tsai, C.-L.; Fu, K.-Y.; Wang, S.-H.; Tseng, H.-J. Characterization and Biocompatibility of Chitosan Nanocomposites. *Colloids Surf. B Biointerfaces* **2011**, *85*, 198–206. [CrossRef]
20. Matica, A.; Menghiu, G.; Ostafe, V. Biodegradability of Chitosan Based Products. *New Front. Chem.* **2017**, *26*, 75–86.
21. Muñoz-Bonilla, A.; Echeverria, C.; Sonseca, Á.; Arrieta, M.P.; Fernández-García, M. Bio-Based Polymers with Antimicrobial Properties towards Sustainable Development. *Materials* **2019**, *12*, 641. [CrossRef]
22. Fernandez-Saiz, P.; Lagarón, J.M.; Ocio, M.J. Optimization of the Film-Forming and Storage Conditions of Chitosan as an Antimicrobial Agent. *J. Agric. Food Chem.* **2009**, *57*, 3298–3307. [CrossRef]
23. Cazón, P.; Vázquez, M. Applications of Chitosan as Food Packaging Materials. In *Sustainable Agriculture Reviews 36: Chitin and Chitosan: Applications in Food, Agriculture, Pharmacy, Medicine and Wastewater Treatment*; Springer: Berlin/Heidelberg, Germany, 2019; pp. 81–123.
24. Ma, Y.; Xu, S.; Yue, P.; Cao, H.; Zou, Y.; Wang, L.; Long, H.; Wu, S.; Ye, Q. Synthesis and Evaluation of Water-Soluble Imidazolium Salt Chitin with Broad-Spectrum Antimicrobial Activity and Excellent Biocompatibility for Infected Wound Healing. *Carbohydr. Polym.* **2023**, *306*, 120575. [CrossRef]
25. Sakai, S.; Yamada, Y.; Zenke, T.; Kawakami, K. Novel Chitosan Derivative Soluble at Neutral PH and In-Situ Gellable via Peroxidase-Catalyzed Enzymatic Reaction. *J. Mater. Chem.* **2009**, *19*, 230–235. [CrossRef]
26. Liu, Y.; Liu, M.; Yang, S.; Luo, B.; Zhou, C. Liquid Crystalline Behaviors of Chitin Nanocrystals and Their Reinforcing Effect on Natural Rubber. *ACS Sustain. Chem. Eng.* **2018**, *6*, 325–336. [CrossRef]
27. Ou, X.; Cai, J.; Tian, J.; Luo, B.; Liu, M. Superamphiphobic Surfaces with Self-Cleaning and Antifouling Properties by Functionalized Chitin Nanocrystals. *ACS Sustain. Chem. Eng.* **2020**, *8*, 6690–6699. [CrossRef]

28. Watthanaphanit, A.; Supaphol, P.; Tamura, H.; Tokura, S.; Rujiravanit, R. Fabrication, Structure, and Properties of Chitin Whisker-reinforced Alginate Nanocomposite Fibers. *J. Appl. Polym. Sci.* **2008**, *110*, 890–899. [CrossRef]
29. Zhong, T.; Wolcott, M.P.; Liu, H.; Wang, J. Developing Chitin Nanocrystals for Flexible Packaging Coatings. *Carbohydr. Polym.* **2019**, *226*, 115276. [CrossRef] [PubMed]
30. Muñoz-Núñez, C.; Hevilla, V.; Zágora, J.; Plachá, D.; Muñoz-Bonilla, A.; Fernández-García, M. Functionalization of Chitosan-Chitin Nanowhiskers Films By Impregnation with Essential Oils Via Supercritical CO_2. *J. Polym. Environ.* **2024**, *33*, 96–111. [CrossRef]
31. Oliveira, F.F.D.; dos Santos, M.R.; Lalli, P.M.; Schmidt, E.M.; Bakuzis, P.; Lapis, A.A.M.; Monteiro, A.L.; Eberlin, M.N.; Neto, B.A.D. Charge-Tagged Acetate Ligands As Mass Spectrometry Probes for Metal Complexes Investigations: Applications in Suzuki and Heck Phosphine-Free Reactions. *J. Org. Chem.* **2011**, *76*, 10140–10147. [CrossRef]
32. Ghosn, B.; Singh, A.; Li, M.; Vlassov, A.V.; Burnett, C.; Puri, N.; Roy, K. Efficient Gene Silencing in Lungs and Liver Using Imidazole-Modified Chitosan as a Nanocarrier for Small Interfering RNA. *Oligonucleotides* **2010**, *20*, 163–172. [CrossRef] [PubMed]
33. Xiao, B.; Wan, Y.; Zhao, M.; Liu, Y.; Zhang, S. Preparation and Characterization of Antimicrobial Chitosan-N-Arginine with Different Degrees of Substitution. *Carbohydr. Polym.* **2011**, *83*, 144–150. [CrossRef]
34. Leceta, I.; Guerrero, P.; De La Caba, K. Functional Properties of Chitosan-Based Films. *Carbohydr. Polym.* **2013**, *93*, 339–346. [CrossRef]
35. Muñoz-Nuñez, C.; Cuervo-Rodríguez, R.; Echeverría, C.; Fernández-García, M.; Muñoz-Bonilla, A. Synthesis and Characterization of Thiazolium Chitosan Derivative with Enhanced Antimicrobial Properties and Its Use as Component of Chitosan Based Films. *Carbohydr. Polym.* **2023**, *302*, 120438. [CrossRef]
36. *ASTM-D1925-70*; Standard Test Method for Yellowness Indez of Plastics. American Society for Testing and Materials: West Conshohocken, PA, USA, 1988.
37. *ASTM E96-00e1*; Standard Test Methods for Water Vapor Transmission of Materials. American Society for Testing and Materials: West Conshohocken, PA, USA, 2000.
38. Garcia, M.A.; Pinotti, A.; Zaritzky, N.E. Physicochemical, Water Vapor Barrier and Mechanical Properties of Corn Starch and Chitosan Composite Films. *Starch-Stärke* **2006**, *58*, 453–463. [CrossRef]
39. Cazón, P.; Morales-Sanchez, E.; Velazquez, G.; Vázquez, M. Measurement of the Water Vapor Permeability of Chitosan Films: A Laboratory Experiment on Food Packaging Materials. *J. Chem. Educ.* **2022**, *99*, 2403–2408. [CrossRef]
40. CLSI. *Methods for Dilution Antimicrobial Susceptibility Tests for Bacteria That Grow Aerobically—Approved Standard*, 9th ed.; CLSI Document M07-A9; Clinical and Laboratory Standards Institute: Wayne, PA, USA, 2012.
41. *ASTM E2149-20*; Standard Test Method for Determining the Antimicrobial Activity of Antimicrobial Agents Under Dynamic Contact Conditions. ASTM International: West Conshohocken, PA, USA, 2020.
42. No, H.K.; Young Park, N.; Ho Lee, S.; Meyers, S.P. Antibacterial Activity of Chitosans and Chitosan Oligomers with Different Molecular Weights. *Int. J. Food Microbiol.* **2002**, *74*, 65–72. [CrossRef] [PubMed]
43. do Amaral Sobral, P.J.; Gebremariam, G.; Drudi, F.; De Aguiar Saldanha Pinheiro, A.C.; Romani, S.; Rocculi, P.; Dalla Rosa, M. Rheological and Viscoelastic Properties of Chitosan Solutions Prepared with Different Chitosan or Acetic Acid Concentrations. *Foods* **2022**, *11*, 2692. [CrossRef] [PubMed]
44. Bezrodnykh, E.A.; Vyshivannaya, O.V.; Berezin, B.B.; Blagodatskikh, I.V.; Tikhonov, V.E. A Walkway from Crayfish to Oligochitosan. *Appl. Sci.* **2023**, *13*, 3360. [CrossRef]
45. Ma, B.; Tan, W.; Zhang, J.; Mi, Y.; Miao, Q.; Guo, Z. Preparation and Characterization of Chitosan Derivatives Bearing Imidazole Ring with Antioxidant, Antibacterial, and Antifungal Activities. *Starch-Stärke* **2023**, *75*, 2200204. [CrossRef]
46. Cui, Y.; Wang, L.; Zhang, Y.; Li, S.; Zhang, F.; Wang, J. Bifunctional Imidazole Cross-Linked Anion Exchange Membrane Based on Quaternized Chitosan with High Performance. *J. Appl. Polym. Sci.* **2025**, *142*, e56786. [CrossRef]
47. Kwon, W.; Jeong, E. Detoxification Properties of Guanidinylated Chitosan Against Chemical Warfare Agents and Its Application to Military Protective Clothing. *Polymers* **2020**, *12*, 1461. [CrossRef] [PubMed]
48. Lustriane, C.; Dwivany, F.M.; Suendo, V.; Reza, M. Effect of Chitosan and Chitosan-Nanoparticles on Post Harvest Quality of Banana Fruits. *J. Plant Biotechnol.* **2018**, *45*, 36–44. [CrossRef]
49. Ma, B.; Zhang, J.; Mi, Y.; Miao, Q.; Tan, W.; Guo, Z. Preparation of Imidazole Acids Grafted Chitosan with Enhanced Antioxidant, Antibacterial and Antitumor Activities. *Carbohydr. Polym.* **2023**, *315*, 120978. [CrossRef]
50. Ibrahim, A.G.; Elgammal, W.E.; Eid, A.M.; Alharbi, M.; Mohamed, A.E.; Alayafi, A.A.M.; Hassan, S.M.; Fouda, A. New Functionalized Chitosan with Thio-Thiadiazole Derivative with Enhanced Inhibition of Pathogenic Bacteria, Plant Threatening Fungi, and Improvement of Seed Germination. *Chemistry* **2023**, *5*, 1722–1744. [CrossRef]
51. Salama, H.E.; Saad, G.R.; Sabaa, M.W. Synthesis, Characterization and Biological Activity of Schiff Bases Based on Chitosan and Arylpyrazole Moiety. *Int. J. Biol. Macromol.* **2015**, *79*, 996–1003. [CrossRef] [PubMed]
52. Wu, L.; Fan, L.; Shi, L.; Wang, C.; Pan, Z.; Xu, C.; Yang, G. Synthesis, Characterization and Antifungal Activity of Imidazole Chitosan Derivatives. *Carbohydr. Res.* **2024**, *544*, 109238. [CrossRef]

53. Song, J.; Feng, H.; Wu, M.; Chen, L.; Xia, W.; Zhang, W. Preparation and Characterization of Arginine-Modified Chitosan/Hydroxypropyl Methylcellose Antibacterial Film. *Int. J. Biol. Macromol.* **2020**, *145*, 750–758. [CrossRef]
54. Kumar, S.; Koh, J. Physiochemical, Optical and Biological Activity of Chitosan-Chromone Derivative for Biomedical Applications. *Int. J. Mol. Sci.* **2012**, *13*, 6102–6116. [CrossRef]
55. Labidi, A.; Salaberria, A.M.; Fernandes, S.C.M.; Labidi, J.; Abderrabba, M. Microwave Assisted Synthesis of Poly(N-Vinylimidazole) Grafted Chitosan as an Effective Adsorbent for Mercury (II) Removal from Aqueous Solution: Equilibrium, Kinetic, Thermodynamics and Regeneration Studies. *J. Dispers. Sci. Technol.* **2020**, *41*, 828–840. [CrossRef]
56. Pestov, A.V.; Mehaev, A.V.; Kodess, M.I.; Ezhikova, M.A.; Azarova, Y.A.; Bratskaya, S.Y. Imidazolyl Derivative of Chitosan with High Substitution Degree: Synthesis, Characterization and Sorption Properties. *Carbohydr. Polym.* **2016**, *138*, 252–258. [CrossRef] [PubMed]
57. Dumitriu, S. *Polysaccharides: Structural Diversity and Functional Versatility*, 2nd ed.; CRC Press: Boca Raton, FL, USA, 2004; ISBN 978-0824754808.
58. Cerrada, M.L.; Sánchez-Chaves, M.; Ruiz, C.; Fernández-García, M. Glycopolymers Resultant from Ethylene-Vinyl Alcohol Copolymers: Degradation and Rheological Behavior in Bulk. *Eur. Polym. J.* **2008**, *44*, 2194–2201. [CrossRef]
59. Sajomsang, W.; Tantayanon, S.; Tangpasuthadol, V.; Thatte, M.; Daly, W.H. Synthesis and Characterization of N-Aryl Chitosan Derivatives. *Int. J. Biol. Macromol.* **2008**, *43*, 79–87. [CrossRef]
60. Islam, M.N.; Khan, M.N.; Mallik, A.K.; Rahman, M.M. Preparation of Bio-Inspired Trimethoxysilyl Group Terminated Poly(1-Vinylimidazole)-Modified-Chitosan Composite for Adsorption of Chromium (VI) Ions. *J. Hazard. Mater.* **2019**, *379*, 120792. [CrossRef]
61. Sadeghpour, A.H.; Yavari, N. Synthesis and Characterization of Chitosan-Based Schiff Base Compounds with Aromatic Substituent Groups. *Biomaterials* **2009**, *3*, 517–524.
62. Caner, H.; Yilmaz, E.; Yilmaz, O. Synthesis, Characterization and Antibacterial Activity of Poly(N-Vinylimidazole) Grafted Chitosan. *Carbohydr. Polym.* **2007**, *69*, 318–325. [CrossRef]
63. Wang, Z.; Yan, Y.; Zhang, Z.; Li, C.; Mei, L.; Hou, R.; Liu, X.; Jiang, H. Effect of Chitosan and Its Water-Soluble Derivatives on Antioxidant Activity. *Polymers* **2024**, *16*, 867. [CrossRef]
64. Ivanova, D.G.; Yaneva, Z.L. Antioxidant Properties and Redox-Modulating Activity of Chitosan and Its Derivatives: Biomaterials with Application in Cancer Therapy. *Biores. Open Access* **2020**, *9*, 64–72. [CrossRef] [PubMed]
65. Abd El-Mawgoud, H.K.; AboulMagd, A.M.; Hemdan, M.M.; Farag, P.S.; Hassaballah, A.I. Synthesis of Novel Imidazole-Based Compounds via Using Sonication Technique: Influence of Chemical Reaction Manipulation, Antiproliferative Activity through Apoptosis Induction, Cell Migration Inhibition, and Antioxidant Properties. *J. Mol. Struct.* **2025**, *1319*, 139611. [CrossRef]
66. Tan, W.; Zhang, J.; Mi, Y.; Li, Q.; Guo, Z. Synthesis and Characterization of α-Lipoic Acid Grafted Chitosan Derivatives with Antioxidant Activity. *React. Funct. Polym.* **2022**, *172*, 105205. [CrossRef]
67. Mi, Y.; Li, Q.; Miao, Q.; Tan, W.; Zhang, J.; Guo, Z. Enhanced Antifungal and Antioxidant Activities of New Chitosan Derivatives Modified with Schiff Base Bearing Benzenoid/Heterocyclic Moieties. *Int. J. Biol. Macromol.* **2022**, *208*, 586–595. [CrossRef]
68. Hasan, A.; Waibhaw, G.; Tiwari, S.; Dharmalingam, K.; Shukla, I.; Pandey, L.M. Fabrication and Characterization of Chitosan, Polyvinylpyrrolidone, and Cellulose Nanowhiskers Nanocomposite Films for Wound Healing Drug Delivery Application. *J. Biomed. Mater. Res. Part A* **2017**, *105*, 2391–2404. [CrossRef]
69. Oliveira, A.C.S.; Ugucioni, J.C.; Borges, S.V. Effect of Glutaraldehyde/Glycerol Ratios on the Properties of Chitosan Films. *J. Food Process. Preserv.* **2021**, *45*, e15060. [CrossRef]
70. Leceta, I.; Guerrero, P.; Ibarburu, I.; Dueñas, M.T.; De La Caba, K. Characterization and Antimicrobial Analysis of Chitosan-Based Films. *J. Food Eng.* **2013**, *116*, 889–899. [CrossRef]
71. Qin, Y.; Zhang, S.; Yu, J.; Yang, J.; Xiong, L.; Sun, Q. Effects of Chitin Nano-Whiskers on the Antibacterial and Physicochemical Properties of Maize Starch Films. *Carbohydr. Polym.* **2016**, *147*, 372–378. [CrossRef] [PubMed]
72. Kadokawa, J.; Takegawa, A.; Mine, S.; Prasad, K. Preparation of Chitin Nanowhiskers Using an Ionic Liquid and Their Composite Materials with Poly(Vinyl Alcohol). *Carbohydr. Polym.* **2011**, *84*, 1408–1412. [CrossRef]
73. Liu, H.; Adhikari, R.; Guo, Q.; Adhikari, B. Preparation and Characterization of Glycerol Plasticized (High-Amylose) Starch–Chitosan Films. *J. Food Eng.* **2013**, *116*, 588–597. [CrossRef]
74. Sriupayo, J.; Supaphol, P.; Blackwell, J.; Rujiravanit, R. Preparation and Characterization of α-Chitin Whisker-Reinforced Chitosan Nanocomposite Films with or without Heat Treatment. *Carbohydr. Polym.* **2005**, *62*, 130–136. [CrossRef]
75. Dubashynskaya, N.V.; Petrova, V.A.; Sgibnev, A.V.; Elokhovskiy, V.Y.; Cherkasova, Y.I.; Skorik, Y.A. Carrageenan/Chitin Nanowhiskers Cryogels for Vaginal Delivery of Metronidazole. *Polymers* **2023**, *15*, 2362. [CrossRef]
76. Rubentheren, V.; Ward, T.A.; Chee, C.Y.; Tang, C.K. Processing and Analysis of Chitosan Nanocomposites Reinforced with Chitin Whiskers and Tannic Acid as a Crosslinker. *Carbohydr. Polym.* **2015**, *115*, 379–387. [CrossRef]
77. Abdel-Monem, R.A.; Rabie, S.T.; El-Sayed, A.A.; Abdelhamid, A.E. Enhanced Performance of Chitosan Film Containing Vinyl Imidazole-Hydroxyethyl Methacrylate Copolymers. *Egypt. J. Chem.* **2024**, *67*, 105–116. [CrossRef]

78. Fernández-Marín, R.; Morales, A.; Erdocia, X.; Iturrondobeitia, M.; Labidi, J.; Lizundia, E. Chitosan-Chitin Nanocrystal Films from Lobster and Spider Crab: Properties and Environmental Sustainability. *ACS Sustain. Chem. Eng.* **2024**, *12*, 10363–10375. [CrossRef]
79. Thomas, M.S.; Koshy, R.R.; Mary, S.K.; Thomas, S.; Pothan, L.A. *Starch, Chitin and Chitosan Based Composites and Nanocomposites*; Springer Briefs in Molecular Science; Springer International Publishing: Cham, Switzerland, 2019; ISBN 978-3-030-03157-2.
80. Holbrey, J.D.; Reichert, W.M.; Swatloski, R.P.; Broker, G.A.; Pitner, W.R.; Seddon, K.R.; Rogers, R.D. Efficient, Halide Free Synthesis of New, Low Cost Ionic Liquids: 1,3-Dialkylimidazolium Salts Containing Methyl- and Ethyl-Sulfate Anions. *Green Chem.* **2002**, *4*, 407–413. [CrossRef]

polymers

MDPI

Article

Air-Assisted Sprayed Flexible Cellulose Acetate/Chitosan Materials for Food Packaging

Nasrin Moshfeghi Far [1], Ana Kramar [2] and Javier González-Benito [1,*]

1 Department of Materials Science and Engineering and Chemical Engineering, Universidad Carlos III de Madrid, Av. Universidad 30, 28911 Leganés, Spain
2 Novel Materials and Nanotechnology Group, Institute of Agrochemistry and Food Technology (IATA), Spanish Council for Scientific Research (CSIC), Calle Catedrático Agustín Escardino Benlloch 7, 46980 Paterna, Spain; akramar@iata.csic.es
* Correspondence: javid@ing.uc3m.es; Tel.: +34-91624948870

Abstract

Cellulose and chitin are the most abundant natural polymers, and their exploitation paves the way for sustainable materials and products. This work investigates the preparation of composites based on acetylated cellulose and partially deacetylated chitin, i.e., chitosan, using versatile and robust air-assisted solution spraying (AASS), a potential method for preparing materials both in situ and ex situ. These materials, in the form of films, despite being prepared from high-molecular-weight and rigid biopolymers, show high flexibility (Young's moduli below 1 GPa), outstanding mechanical properties (tensile strengths above 19 MPa and strain at failure higher than 2%), and bioactivity towards *E. coli*. The unprecedented flexibility, obtained without the use of any plasticizer or by casting with humidity control, is a direct consequence of the specific film morphology, whereby films are constituted from merging droplets. Depending on the solution properties (viscosity, surface tension), various droplet sizes are obtained, thus influencing the roughness and indirectly the wettability. Wettability analysis towards water and oil revealed higher contact angles towards both fluids as the content of chitosan increases in the composite what directly impacts packaging applications by better protecting the food. Besides this, higher chitosan content in the composite (7.5% w/w) enabled bioactivity against *E. coli*, where colony development was inhibited on the film surface compared with the neat cellulose acetate. This study shows a very high potential for AASS for obtaining uniform thin flexible films for food packaging applications, allowing faster drying and lower energy consumption than other film-forming techniques.

Keywords: food packaging; cellulose acetate; chitosan; air-assisted solution spraying; solution blow spinning; multifunctional materials

1. Introduction

There is growing awareness in society of the fundamental danger lying in the non-biodegradability of plastic materials, which accumulates in the environment, causing negative impacts [1]. As a result, the use of sustainable bioplastics has gained significant interest, mainly in sectors such as food packaging. Among these materials, cellulose acetate (CA) is receiving special attention because it arises from cellulose, which is one of the most abundant polymers on Earth, and it has a high ability for film and fiber formation [2,3]. Furthermore, CA can have interesting properties when it is constituted by nanofibers, particularly with the advent of electrospinning. Among others, cellulose acetate (CA) nanofibers find several applications, including tissue engineering, filtration membranes, and wound dressings [4–6]. In the particular case of food packaging, electrospinning and solution casting have been recently revisited for CA, allowing the preparation of thin materials suitable for use as polymeric films for food packaging [7,8]. Besides its natural origin, being the derivative of cellulose, cellulose acetate is convenient from a circular bioeconomy aspect, as it is biodegradable as well. Although without directly using the ASTM D6400 specification Juergen Puls et al. used the experimental conditions (aerobic composting and CO_2 evaluation using standard methods) to report that CA films with a DS = 2.57 require about 18 days to fully degrade under composting conditions, using a Sturm-type test (based on CO_2 evolution measurement) [9]. CA of lower DS (below 2) is readily biodegradable under appropriate composting conditions, but is also relatively hydrophilic. Therefore, it is important to find adequate biodegradability, a balance between hydrophilicity and hydrophobicity, while maintaining good mechanical properties and flexibility at the same time, for food packaging applications.

However, achieving this balance of properties can be challenging, especially when using hydrophilic and rigid polymers like polysaccharides, such as cellulose acetate (CA). Their rigidity, due to their intra- and intermolecular interactions (especially hydrogen bonds in partially hydroxylated regions), can lead to brittleness if they are not conveniently modified with plasticizers or prepared under specific relative humidity [10]. Food quality and safety quickly deteriorate when made from hydrophilic materials because their water barrier properties are poor [11]. Although CA film exhibits a static water contact angle of approximately 60°, indicating that it is hydrophilic, but less so than pure cellulose [12], representing a little improvement with respect to neat cellulose. However, it would be interesting to achieve an even lower hydrophilicity. In this sense, several approaches based on modifications with other polymers and particles have been tried, for example, using polyvinylidene fluoride (PVDF) [13] or PVDF with poly methylmethacrylate (PMMA) [14] in the case of polymers, and among others, SiO_2, ZnO as nanoparticles [15]. Besides modifying CA with other components, sometimes a better way to tune hydrophobicity is to tailor the morphology of the materials (size and shape of nano or microfeatures, pores, for instance) as has been demonstrated by the use of electrospinning to achieve hydrophobic and even superhydrophobic properties [16,17]. It is clear, therefore, that many factors influence the final wettability behavior of these materials, and further research should be conducted to shed more light on this issue. In terms of the optimization of morphology, processing conditions constitute a very important issue to take into account. Electrospinning (ES) is perhaps the most well-known and currently used technique to obtain complex morphologies from continuous films to nanofibrous polymer-based materials [18]. However, some drawbacks can be considered for its use; for example, a high electric field is required, there is a limitation in the choice of solvents for polymers to those that interact properly with the electric field, and the production rates are relatively low. For this reason, other techniques should be explored to attempt to overcome the aforementioned drawbacks. Among possible techniques, solution blow spinning (SBS), or gas-assisted solution spraying when fibers are not obtained, is being studied to address those limitations [19,20]. In terms of morphologies, SBS offers results comparable to electrospinning but with a faster production rate, using a pressurized gas instead of an electric field as the driving force to eject and stretch the solution and deposit the polymer-based material onto a substrate. In addition, SBS is industrially scalable by using multi-nozzle or continuous spray heads mounted on automated lines where the so-produced atomized or spun polymer must be deposited onto a moving substrate or conveyor belt [21].

In addition, SBS, or more conveniently and from now on called air-assisted solution spraying (AASS), when air is used as the gas and no fibers are obtained, was already used to prepare films for different purposes [22,23], which reinforces the idea of preparing CA-based materials through the use of SBS. Among the advantages offered by AASS, it can be highlighted that there is the possibility of dispensing the materials over any kind of substrate, allowing the "in situ" preparation of materials. Considering the above-mentioned, this processing method would be highly versatile since it allows those materials to be directly sprayed on foods, fruits, for instance, or simply deposited on a substrate from which they would be extracted for other purposes, such as the preparation of functional sachets for food packaging applications.

However, AASS is proving challenging for the processing of CA [22], because of the difficulty in producing very particular morphologies such as defect-free nanofibers. CA spinnability by AASS can be facilitated by the use of other polymers included in the spun solution. According to Claro and coworkers, poly(ethylene oxide) (PEO) was used together with CA to finally produce fibrous CA-based materials by AASS in the form of flexible ionic conductive cellulose mats [24]. It was also successfully produced CA fibers using PEO as a spinning aid, showing the importance of the molecular weight of polymers and solvents [22]. By co-spinning CA nanofibers with polyacrylonitrile, Dadol and coauthors also successfully produced CA nanofibers [25]. Therefore, still more efforts should be focused on finding easier AASS conditions to prepare nanofibrous CA-based materials. Therefore, looking for adequate processing conditions and a solution system, nearly any morphology should be able to be obtained.

On the other hand, plastic materials for food packaging should possess additional properties such as mechanical consistency, chemical resistance, thermal stability, and environmental friendliness, most of which are already provided by CA. However, it is highly recommended to incorporate special functionalities, such as antimicrobial activity and the ability to delay oxidation processes.

The most common way to impart antibacterial activity to plastics for food packaging is by the incorporation of antimicrobial agents, such as natural organic compounds like nisin [26], chitosan [27], and essential oils [28] or inorganic nanoparticles like silver (Ag) nanoparticles [29], zinc oxide (ZnO) nanoparticles [30] and titanium dioxide (TiO_2) nanoparticles [31].

Among them, chitosan (CS) is currently receiving special attention due to multiple reasons that make it very attractive for food packaging purposes. CS has a natural origin, is biodegradable, and non-toxic. CS is derived from chitin that can be extracted from shrimps and crab shells, making it a sustainable and eco-friendly material. CS is recognized as GRAS (Generally Recognized as Safe) by the FDA for food applications, without releasing harmful residues. In addition, CS has a broad spectrum for antimicrobial activity since it is effective against Gram-positive and Gram-negative bacteria, fungi [32], and even some viruses [33,34]; however, the antiviral effectiveness of chitosan is still not well understood and depends on factors such as pH, charge, molecular weight, etc. CS can enhance oxygen and moisture barrier properties, prolonging shelf life [35,36], and finally, CS has shown great potential for controlled release processes of antimicrobial compounds to extend food preservation [37]. The antimicrobial activity of chitosan itself mostly depends on the degree of substitution (deacetylation) and molecular weight.

The study focuses on optimizing the air-assisted solution spraying (AASS) as a potential sustainable, effective, and low-cost processing method to prepare new materials for food packaging. These materials are prepared in the form of films mainly constituted by cellulose acetate (CA) with improved properties in terms of flexibility, antimicrobial action, and hydrophobicity, for a better performance as a food packaging material. In this sense, a full characterization of the films' morphology, mechanical strength, and wettability, together with the evaluation of their antibacterial activity against E. coli, is carried out.

To tailor the surface properties and enhance the final performance of materials for food packaging and enhance food shelf life, CA is modified by the addition of chitosan (CS), and the morphology of the materials is controlled by selecting appropriate conditions of the AASS processing method (pressurized air flow, solution feeding rate, working distance, and composition of polymer solutions) for preparation of thin film-like materials from a polymeric blend. Therefore, this work aims to develop innovative films from cellulose acetate and chitosan using AASS, without plasticizers, suitable for food packaging applications and aligned with environmental sustainability goals.

2. Materials and Methods

2.1. Materials

Cellulose acetate, CA (Sigma Aldrich, St. Louis, MO, USA, average Mn ~ 30,000 g/mol, acetyl content 39.8 wt%, degree of substitution, DS = 2.45 that represents the average number of acetate groups per anhydroglucose unit) and chitosan, CS, (Sigma-Aldrich, St. Louis, MO, USA, low molecular weight, zero-shear viscosity measured using Haake Viscotester IQ–Thermo Fisher Scientific (Waltham, MA, USA), at 25 °C of 1 wt% solution in 2% acetic acid is 1.73 Pa·s, degree of deacetylation, DDA = 66%, according to the method described elsewhere [38]) were used as received. After a deep solubility study (see Supplementary Materials), it was found that the best solvent systems for this set of polymers were formic acid, FA (Panreac, Barcelona, Spain, 85% purity), and acetic acid, HAc (Panreac, Barcelona, Spain, glacial 91% purity). The use of 91% acetic acid is intended to improve the spray processing of the corresponding polymer solution. Since the solvents are not involved in a chemical synthesis and are ultimately removed during the AASS preparation of these materials, the choice of 91% acetic acid is considered acceptable, particularly as it is expected to facilitate solvent evaporation during the AASS process.

As a pH indicator for the titration of CA to obtain the degree of substitution DS, Phenolphthalein (Panreac, Barcelona, Spain, 318.33 g/mol) was used. On the other hand, methyl orange (Panreac, Barcelona, Spain, 327.34 g/mol) was used as the pH indicator for the titration of CS to obtain its degree of deacetylation, DDA. Ethanol, EtOH (Sigma Aldrich, St. Louis, MO, USA, 96% purity), was used to dissolve CA. Aqueous solution of sodium hydroxide 1.0 M, NaOH (Sigma Aldrich, St. Louis, MO, USA), was used for saponification of CA and back-titration of CS, while standardized hydrochloric acid solution 1.0 M, HCl (Sigma Aldrich, St. Louis, MO, USA), was used for back-titration of CA and initial dissolution of CS.

2.2. Degree of Substitution (DS) of CA

To understand the properties and behavior of materials prepared in this work, information about the degree of substitution (DS) of CA is required, or the average number of acetyl groups substituting the hydroxyl groups within the glycosidic units. A typical method of saponification with subsequent back-titration was carried out. The process was performed in triplicate at room temperature, with continuous magnetic stirring for 60 min to obtain both the degree of acetylation (%AG) and DS by the use of Equations (1) and (2) that were adapted from Samios et al. [38], where the factor 43 g/mol refers to the molar mass of the acetyl group (CH_3CO–) and the value 102.40 represents the molar mass of the original repeating unit (before acetylation) used in the calculation of substituents.

$$\%AG = \frac{\left[\left(V_{b_i} - V_{b_t}\right) \times C_b - V_a \times C_a\right] \times 43 \times 100}{m_{CA}} \quad (1)$$

$$DS = \frac{3.86 \times \%AG}{102.40 - \%AG} \quad (2)$$

where V_{b_i} is the volume of NaOH added to the system, V_{b_t} is the volume of NaOH utilized for titration, C_b is the molar concentration of the base, V_a is the volume of HCl introduced into the system, C_a is the molar concentration of the acid, and m_{CA} is the mass of CA used.

2.3. Degree of Deacetylation (DDA) of CS

Similarly to CA, the degree of deacetylation of CS is required to better understand the final properties. Also, in this case, an acid-base titration was used following the method described in [39] and using Equation (3).

$$DDA = \left[1 - \frac{(V_b \times C_b \times 161)}{(m_{CS} \times 1000)}\right] \times 100 \tag{3}$$

where V_b is the volume of NaOH (in mL) used for the titration, C_b is the concentration of NaOH used in mol/L, 161 is the molar mass of the glucosamine monomer unit ($C_6H_{11}NO_4$), in g/mol, and m_{CS} is the weight in grams of CS used.

2.4. Sample Preparation

Before choosing the best conditions to obtain solutions where both CA and CS are compatible with a solvent system, a deep solubility study was performed (see Supplementary Materials). Solvents and solubility conditions were chosen considering the most homogeneous stable solutions in terms of clarity or liquid transparency, for which no phase separation was considered.

First, the polymer solutions to be blow spun were prepared by dissolving the polymer systems in both FA and HAc. CA was added to the corresponding solvent to reach a concentration of 8% by weight. Then, specific amounts of CS were added to prepare the final solid materials with different compositions for the blend CA/CS, 0%, 2.5%, 5.0% and 7.5% by weight in chitosan (percentages correspond to the relative amount of chitosan with respect to the whole blend or material), which were labeled as CA, CA/CS-1; CA/CS-2 and CA/CS-3, respectively, adding the prefix HAc- or FA- depending on the solvent used.

Once the solutions are prepared, they are subjected to AASS using a home-designed and made automated device [40]. Through the use of an automatic pump and a syringe, solutions are injected in a coaxial nozzle with an inner channel of 0.6 mm diameter protruding 2 mm from the nozzle. On the other hand, pressurized air is passed through the outer nozzle of 1.0 mm, exerting the required force on the solution to spray it to the collector while favoring the evaporation of solvent. The materials were collected on a rotating drum (rotational speed 250 rpm). When the solvent used was HAc, AASS processing conditions were 12 cm of working distance (WD) or length between the collector and the inner nozzle, solution injection rate of 0.125 mL/min, and air pressure of 1.0 bar. When the solvent used was FA, AASS processing conditions were 15 cm of WD, a solution injection rate of 0.125 mL/min, and an air pressure of 1.5 bar. The AASS conditions, chosen according to the solvent, were selected simply through several trial-and-error tests until those that ensured the formation of a homogeneous film in terms of thickness and prevented nozzle clogging during the AASS process were identified. As an example, in Figure S1, representative images of films obtained by spraying are shown (regardless of the composition and the solvent used, all films are very similar), for which similar thicknesses were obtained, ~300 μm.

2.5. Methods and Equipment

2.5.1. Viscosity

Viscosity measurements were performed using a Haake iQ viscometer. Solutions or suspensions were placed between two circular plates at constant temperature (25 °C), and subjected to an oscillatory shear force, in the frequency range 100–1000 s^{-1}, with an acquisition time of 30 s per measurement (100 points). In Table 1, the values of viscosities in Pa·s obtained at 0.1 kHz for all systems to be sprayed are collected.

Table 1. Values of viscosities and surface tensions of all solutions prepared to be sprayed are collected.

System	Viscosity (Pa·s)	Surface Tension ($mN \cdot m^{-1}$)
HAc-CA	0.379 ± 0.031	40.2 ± 2.7
HAc-CA/CS-1	0.895 ± 0.073	41.5 ± 2.4
HAc-CA/CS-2	1.096 ± 0.086	43.6 ± 2.4
HAc-CA/CS-3	1.873 ± 0.145	45.6 ± 1.8
FA-CA	0.239 ± 0.021	45.0 ± 2.2
FA-CA/CS-1	0.567 ± 0.046	45.6 ± 2.3
FA-CA/CS-2	0.876 ± 0.071	47.6 ± 2.3
FA-CA/CS-3	1.328 ± 0.109	50.0 ± 2.7

2.5.2. Surface Tension

Surface tension was determined using the pendant drop method with a tensiometer OCA-15 Plus Goniometer (Data Physics, Neuterk Instruments, Eibar, Spain). A standard needle with a diameter of 1.65 mm was used, and at least 10 drops were measured per sample, with photographs taken 30 s after the drop formation at room temperature. Finally, from the pear-shaped drop image analysis and using the Young-Laplace equation, the surface tension was obtained. In Table 1, the values of surface tensions in $mN \cdot m^{-1}$ of all systems prepared to be sprayed are collected.

One-way ANOVA test showed there is a statistically significant difference (with α level 0.05) in the values of viscosity and surface tension between the solutions that contain different amounts of chitosan.

2.5.3. Structural Characterization

Attenuated Total Reflectance Fourier transform infrared spectroscopy (ATR-FTIR) was used to analyze the chemical structure of materials. A Nicolet iS5 spectrometer (Thermos Scientific, Thermofisher, Madison, WI, USA) and an attenuated total reflectance (ATR) device with a diamond window GladiATR (PIKE Technologies, Madison, WI, USA) were utilized. To obtain the spectra, 32 scans at a resolution of 4 cm^{-1} within the range 400–4000 cm^{-1} were performed with a Golden Gate ATR accessory to create Fourier-transformed infrared spectra.

On the other hand, X-ray diffraction (XRD) was also used to study the structure of the materials. XRD patterns were obtained using an X-ray powder diffractometer Bruker ECO D8 Advance (Bruker, Karlsruhe, Germany) from the interaction with the Cu-Kα1 radiation (λ = 1.5418 Å) in the range 5° to 50° in 2θ, with a scanning rate of 2 s per step and a step size of 0.02°.

2.5.4. Morphological Characterization

The morphology of the materials was inspected by field emission scanning electron microscopy (FESEM) using a TENEO-FEI field emission scanning electron microscope. FESEM images were generated from secondary electrons from a primary electron beam accelerated with a 10 kV potential and detected with an ETD detector. Samples were mounted onto SEM stubs using double-sided carbon tape and then sputter-coated with gold for 90 s using a Leica EM ACE200 low-vacuum coater (Leica Microsystems, Wetzlar, Germany) to render them conductive, preventing both electrostatic charge buildup and localized heating. To analyze images and determine different morphological parameters, the free software ImageJ (V.1.52a, National Institutes of Health, Bethesda, MD, USA) was used.

The morphology was also studied using an Olympus DSX500 optical profilometer (Olympus Iberia, Barcelona, Spain). In this case, images were used, among other things, to determine roughness parameters using the cut-off wavelength, λ_c, or the length scale used to filter out surface features larger or smaller than it when measuring roughness, ensuring that only the relevant roughness is analyzed according to ISO 4288 [41].

Regardless of the technique used to study the morphology, only the surface of the samples directly in contact with air during their preparation by AASS was inspected.

To determine the porosity, the well-known gravimetric method was used [42]. Specimens of samples were weighed out in an analytical balance to obtain the mass, m, and dimensioned using an Easy-check Neurtek Instrument with an accuracy of ±1 µm to measure thicknesses, h, and a digital caliper to measure widths, w, and lengths, l. In this way, volumes of specimens were obtained as $h \times w \times l$. Finally, the specimen density was obtained by dividing the mass by the volume. After obtaining the densities of three specimens, the final sample density was calculated as an average (ρ_s). The porosity was finally determined following Equation (4), where theoretical density (ρ_t) is included, and that was estimated from a simple rule of mixtures (Equation (5)).

$$\phi = 1 - \frac{\rho_s}{\rho_t} \tag{4}$$

$$\rho_t = \rho_{CA} \cdot \phi_{CA} + \rho_{CS} \cdot \phi_{CS} \tag{5}$$

where ρ_{AC} and ρ_{CS} are the densities of the polymers, cellulose acetate (~1.3 g·cm^{-3} [43]) and chitosan (~1.5 g·cm^{-3} based on the data reported by D. Lupa et al. [44]), respectively, and ϕ_{AC} and ϕ_{CS} are the volume fractions in the corresponding sample.

2.5.5. Water Vapor Permeability

To study the water vapor permeability, the water vapor transmission rate (WVTR) was determined using the method provided in the standard ISO 2528:2017 [45], slightly modified to test the kind of specimens prepared in this work. The experiments were conducted in triplicate, and the mass of water per square meter of film per day was utilized to compute the permeability.

2.5.6. Solvophilicity Behavior

Water and oil contact angle measurements were performed in an OCA-15 Plus Goniometer (Data Physics, Neurtek Instruments, Eibar, Spain) through the sessile drop contact angle method. Photographs of 3 µL volume of distilled and deionized water or oil (extra virgin olive oil) drops deposited on the films were captured immediately after their deposition. After that, image analysis was carried out to obtain the angle formed between the droplet and the surface of the material. At least five measured angles were used to finally obtain the contact angle as an average. It is important to highlight that surface tension (mN·m^{-1}) and surface energy (mJ·m^{-2}) are numerically equivalent and represent the same quantity.

2.5.7. Thermal Behavior

Thermal relaxations of the films were studied by differential scanning calorimetry (DSC) using a Mettler Toledo 822e calorimeter under a nitrogen atmosphere. Samples (4–6 mg) were weighed in aluminum pans sealed with perforated covers to keep constant pressure. After that, samples were subjected to three thermal cycles of heating-cooling-heating, only the first heating was used to compare the thermal behaviors of the different materials, since information from the prepared materials is required for the present work. This first heating cycle consisted of a ramp of temperatures from 50 °C to 260 °C at 10 °C/min. Glass transition and melting temperatures were extracted from the thermograms, and the crystalline fraction of the materials, χ, was calculated using Equation (6) from the area of the endothermic peak and dividing by the corresponding enthalpy of fusion of triacetate cellulose 100% crystalline, $\Delta H_{100} = 58.8\ \text{J}\cdot\text{g}^{-1}$ [46].

$$\chi = \frac{\Delta H_m \cdot (1 - x)}{\Delta H_{100}} \tag{6}$$

where ΔH_m is the enthalpy of fusion obtained from the area under the endotherm of the first heating cycle, and x is the weight fraction of chitosan.

2.5.8. Mechanical Behavior

Tensile tests of samples were carried out using a universal testing machine, Microtest DT/005/FR (Microtest S.A., Madrid, Spain) with a load cell of 50 N. Rectangular specimens (40 mm length, 10 mm width, and ~0.02 mm thickness) were tested, using a gauge length of 20 mm and applying a loading rate of 5 mm/min. At least 5 specimens (cut following the direction of collector rotation) of each sample were tested to obtain, from the analysis of the corresponding stress–strain curves, the mechanical parameters as mean values. Before each test, specimen thickness was measured using a Digimatic micrometer (Mitutoyo Corporation, Barcelona, Spain) with ±1 μm accuracy.

2.5.9. Bioactivity Assay

To ensure the absence of microorganisms, all materials were sterilized by exposure to ultraviolet (UV) light for a minimum of 20 min before use. A commonly used Escherichia coli strain, DH5α, was selected for bacterial culturing aligned with methods outlined in ISO 22196 [47] and ASTM E2149 [48]. The idea of using a DH5α strain is to begin with a preliminary and straightforward study of the antimicrobial activity of the materials under investigation, more specifically, the effect of the presence of chitosan. To prepare the bacterial suspension, 90 μL of bacterial culture was mixed with 2910 μL of Luria–Bertani (LB) medium, resulting in a total volume of 3 mL. The mixture was incubated overnight at 37 °C with continuous shaking to promote bacterial growth. On the second day, the overnight bacterial culture was diluted 1:100 in fresh LB medium. Subsequently, 1 mL of this diluted bacterial suspension was added to each test material. The samples were then incubated at 37 °C for 3 h with shaking to allow bacterial adhesion. Following the incubation period, all liquid was removed, and the samples were washed by adding 1 mL of physiological saline solution (0.9% NaCl). The solution was gently agitated and subsequently discarded to remove excess bacteria. To collect the bacteria adhered to the surface of the test materials, sterile cotton swabs were used. The swabs were gently pressed against the surface of the samples to ensure bacterial collection, and then they were immersed in fresh LB medium. This mixture was incubated at 37 °C for 1 h to allow bacterial resuspension in the liquid. After incubation, serial dilution and plating were followed. The cotton swabs were removed, leaving a liquid containing the collected bacteria. Serial dilutions were performed in a stepwise manner: 1/100 (10^{-2}); 1/1000 (10^{-3}); 1/10,000 (10^{-4}) and 1/100,000 (10^{-5}). An LB-agar plate was divided into four sections, and 10 μL of each dilution was carefully placed onto the corresponding section. The plate was incubated overnight at 37 °C in an inverted position to facilitate bacterial colony formation. This method allows for the evaluation of bacterial attachment and viability on different materials, providing a quantitative assessment of their antibacterial properties.

2.5.10. Statistical Analysis

When enough information was available to conclude, to determine whether there are statistically significant differences between the means of eight independent groups of data (associated with parameters of characterization) corresponding to eight samples studied, One-Way ANOVA analysis of variance was performed with a significance level of 0.05.

3. Results

3.1. Structure of the Materials

XRD analysis was performed to study possible changes in the polymer structure as a result of the AASS process and interactions between polymers in the blends. Figure S2 depicts the X-ray diffraction patterns of the as-received polymers CA and CS, respectively. In the case of as-received CA, three sharp peaks at 9°, 10°, and 13° are observed, which correspond to residual crystalline domains from cellulose [49]. On the other hand, peaks at 17°, 21°, and 23° of crystallographic planes (110), (020), and (002) reveal the presence of typical cellulose acetate [50]. When analyzing the CS diffractogram, it is observed that the typical peaks at 10° and 20° are associated with the (020) and (110) planes [51].

In Figure 1, the XRD patterns of the different systems prepared by SBS based on CA are shown. Pure chitosan could not be prepared by AASS because of the difficulties of nozzle clogging. As can be seen, regardless of the solvent used to dissolve the polymers (HAc and FA), the highly ordered structure of the as-received polymers was lost, being ever more evident when chitosan is added, as can be deduced from the relative decrease in intensity of the broad peak appearing at $2\theta = 8.5°$. Therefore, as a conclusion of these results, it can be said that AASS processing of CA decreases its crystallinity, being further promoted by the presence of CS.

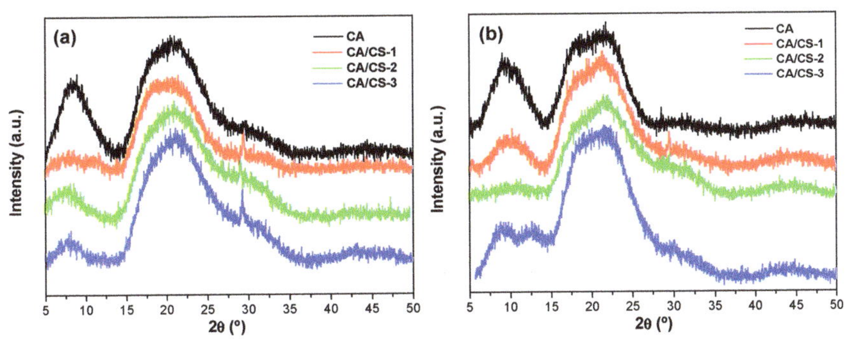

Figure 1. XRD patterns of the different systems prepared by AASS based on CA. (**a**) Acetic acid (HAc) and (**b**) Formic acid (FA) used as solvents.

In order to find more information about not only the appearance of crystals, but also specific interactions between the polymers in the CA/CS blends, ATR-FTIR was utilized. Typical peaks associated with cellulose acetate were discernible across all spectra (Figure 2). The typical bands of cellulose acetate at 1737 cm^{-1} (carbonyl, C=O, stretching of the acetyl groups), at 1220 cm^{-1} (acetate ester groups, C–O), and 1030 cm^{-1} (C–O–C stretching of glycosidic linkage in cellulose backbone) [52] do not experience any change with the solvent used for AASS nor with the addition of chitosan. In addition, barely noticeable, the broad O–H stretching band at ~3470–3550 cm^{-1} corresponds to residual hydroxyl groups, which points out a high degree of substitution. With respect to chitosan, considering its low concentration in the blends, it is expected that its characteristic absorption bands will be overlapped by the CA bands. In fact, identification of specific chitosan bands, such as the

N–H amine stretching band, was not possible. Upon the addition of CS significant shifts or changes in CA bands were not detected which suggests limited chemical interaction The only small difference observable in the materials obtained with both solvents is the relative increment of a small peak at 875 cm^{-1} concerning the peak at 899 cm^{-1} that can be assigned to the cellulose ring vibrations (C–O–C and C–H bending). The slight variation can only be attributed to a minor change in the crystallinity of CA, which may result from a simple nucleating effect induced by the presence of CS. In order to confirm this last observation, DSC was carried out to identify thermal transitions and determine the degree of crystallinity.

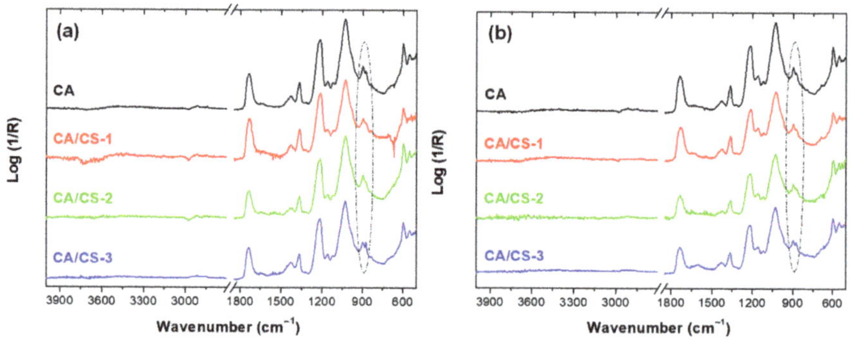

Figure 2. Normalized FTIR spectra of the different systems prepared by AASS based on CA. (**a**) Acetic acid (HAc) and (**b**) Formic acid (FA) used as solvents.

Here, it is convenient to highlight that the presence of acetic acid or formic acid in plastics can lead to rejection for food packaging applications by the FDA, depending on their concentration, purpose, and potential migration into food. However, in principle, the idea of using AASS is to achieve a material completely free of solvents. This objective is usually fulfilled thanks to the intense work performed looking for the proper processing conditions, which, for this particular issue, are related to the air pressure and the working distance. One way to ensure the absence of solvents in the final materials is through information delivered by FTIR. As can be seen in Figure 2, there is no evidence of HAc or FA, since the intense peaks corresponding to the O-H stretching vibrations that must be found in a broad band from 3200 to 3550 cm^{-1} are not observed. Therefore, in the case of the presence of these two solvents, they should be in the form of traces below the sensitivity limit of the spectrophotometer, which is around 0.5%. However, before validating the materials for their final use, a migration test using food simulants should be conducted.

In Figure 3, complex DSC thermograms are observed for the AASS CA-based materials prepared in the present work, where several endo- and exothermal phenomena are shown. This particular complex thermal behavior is found in other cellulose acetate systems [53,54], which usually has led to contradictory discussions. Due to this, the assignment should be made, being supported by results obtained by other techniques, such as DMA. Following this, and based on the results obtained by Scandola et al. [53,55], it seems more reasonable to assign the glass transition, or more accurately, the α relaxation, to the heat capacity increase observed at around 195 °C (Figure 3); however, it should ideally be confirmed by complementary techniques, such as DMA. Finally, there is a clear endothermic peak at 225 °C associated with the melting process of CA [56]. Finally, although there is some evidence in the bibliography of a small endothermic peak for chitosan between 130 °C and 150 °C, which is assigned to the relaxation of chitosan chains rather than melting [57], in our samples, it is not expected any clear signal because of the small amount of chitosan in the whole sample (maximum 7.5% wt), which would mean about 0.4 mg in the analyzed sample. In Table 2, values of glass transition temperature, T_g, melting temperature, T_m, and degree of crystallinity, χ, are gathered for the samples of CA-based materials.

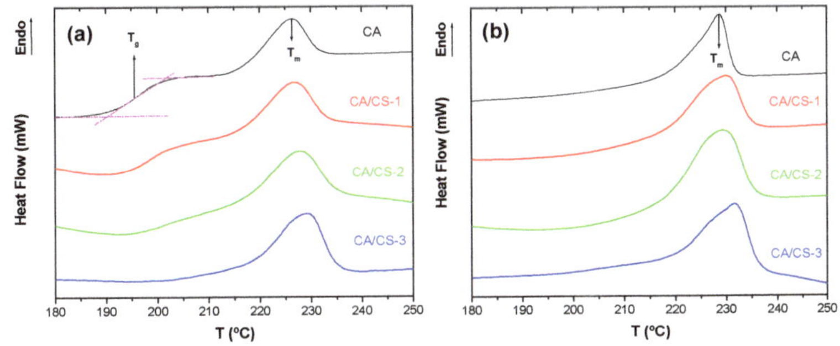

Figure 3. DSC thermograms are observed for the AASS CA-based materials prepared in the present work. (**a**) Acetic acid (HAc) and (**b**) Formic acid (FA) used as solvents.

Table 2. Values of glass transition temperature, T_g, melting temperature, T_m, and degree of crystallinity, χ, of the CA-based materials.

Sample	T_g (°C)	T_m (°C)	χ (%)
HAc-CA	197	226	6.0
HAc-CA/CS-1	198	227	6.5
HAc-CA/CS-2	198	228	5.0
HAc-CA/CS-3	-	229	7.5
FA-CA	-	229	15.3
FA-CA/CS-1	-	229	13.6
FA-CA/CS-2	-	229	11.5
FA-CA/CS-3	-	231	11.2

As can be seen, only when samples are prepared using solutions made in acetic acid, both glass transition and fusion of CA are observed, pointing out that more crystalline samples are obtained when formic acid is used as solvent. In fact, in Table 2 values of crystallinity degree reflect that samples prepared with formic acid as solvent present around 65% higher crystallinity. A quite lower fraction of the amorphous phase must be the reason the calorimeter is not sensitive enough to show the change in the heat capacity associated with the glass transition. One possible reason for this result is the possibility of higher deacetylation of CA when it is dissolved in FA. In fact, on average, the ratio between the absorbances at 1030 cm^{-1} (C–O–C stretching of glycosidic linkage in cellulose backbone) and at 1220 cm^{-1} (acetate ester groups, C–O) is higher when acetic acid is used as the solvent (0.8 > 0.7). It is known that the higher the degree of substitution (DS) in cellulose, the lower the crystallinity. For example, cellulose acetate with a DS of 2.5 (coincident with our as-received CA) exhibits a crystallinity index close to 10% (similar to our results when CA-based materials are obtained from a solution made with HAc, Table 2), and the lower the DS, the higher the crystallinity [58]. Here, it is proposed that formic acid, being a protic and relatively strong organic acid, can act as a catalyst for the hydrolysis of the ester bonds in cellulose acetate [59]. This results in the cleavage of acetyl groups (-$COCH_3$) and the regeneration of hydroxyl (-OH) groups on the cellulose backbone. However, the above reason should be taken with caution because stronger chemical evidence would be necessary, such as direct chemical analysis for the determination of the degree of substitution.

On the other hand, two behaviors are observed as a function of the concentration of CS, depending on the solvent used to prepare the solutions to be blow spun. When HAc is used as a solvent, there is an increase in crystallinity as the concentration of CS increases. This result can be explained through the consideration of a nucleation effect since CS is less soluble in HAc, so it is expected to have small CS domains within the CA matrix while the latter is solidifying because of solvent evaporation. On the other hand, when FA is used as a solvent, there is a slight decrease in the crystallinity as the concentration of CS increases. This result can be explained by partial neutralization of FA because of the presence of the amino groups, which should shift the hydrolysis of CA to the reactants or lower deacetylation. In addition, FA is a better solvent for CS than HAc.

3.2. Morphology

Optical profilometer and scanning electron microscopy were the microscopy techniques chosen to study the morphology of the sprayed CA-based samples. The analysis of the morphology of samples is crucial to understand several properties, such as solvophilicity or the ability of a solvent to impregnate a surface. In Figure 4, topographical images obtained with the profilometer are presented.

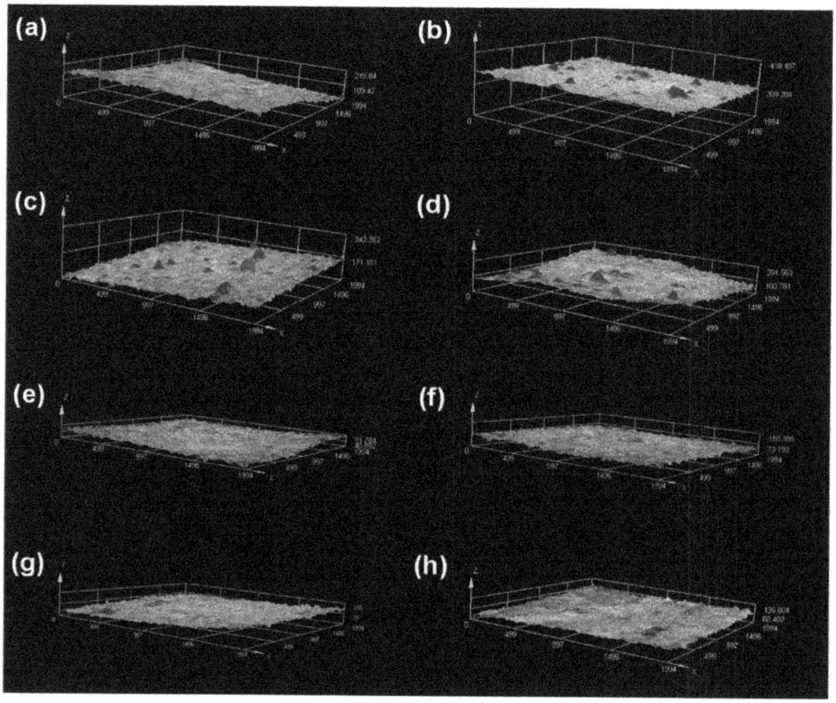

Figure 4. Representative 3D images obtained with an optical profilometer of the surfaces of cellulose acetate/chitosan films prepared by AASS (axes units, μm). (**a**) HAc-CA; (**b**) HAc-CA/CS-1; (**c**) HAc-CA/CS-2; (**d**) HAc-CA/CS-3; (**e**) FA-CA; (**f**) FA-CA/CS-1; (**g**) FA-CA/CS-2; (**h**) FA-CA/CS-3.

As can be seen, when using acetic acid as the solvent, more heterogeneous surfaces were obtained for the CA-based materials, where randomly dispersed surface small protuberances are observed over the entire surface. In addition, it seems that the higher the concentration of CS, the more heterogeneous the topography. However, it is also interesting to have quantitative topographical information from the roughness parameters. In Table 3, the values of three roughness parameters (R_a, R_q, and Rz, see experimental section) are gathered. As can be seen, the quantitative results obtained through roughness confirm what is observed in the images. It seems, therefore, that the appearance of those protuberances causes the roughness variations. One-way ANOVA test showed there is a statistically significant difference (with α level 0.05) in the roughness represented by the parameters given in Table 3 when comparing the samples as a function of the concentration of chitosan.

Table 3. The roughness parameters of the CA-based materials obtained by AASS.

Sample	R_a (μm)	R_q (μm)	R_z (μm)
HAc-CA	4.4 ± 0.7	5.9 ± 1.1	34.4 ± 9.5
HAc-CA/CS-1	5.5 ± 2.0	7.0 ± 2.4	36.1 ± 8.7
HAc-CA/CS-2	6.7 ± 2.1	8.5 ± 2.7	41.1 ± 16.4
HAc-CA/CS-3	10.0 ± 2.2	12.4 ± 2.2	58.8 ± 10.6
FA-CA	4.2 ± 0.8	5.4 ± 1.0	29.2 ± 6.1
FA-CA/CS-1	4.5 ± 0.6	5.7 ± 0.9	29.4 ± 7.0
FA-CA/CS-2	4.7 ± 0.8	5.9 ± 1.0	29.5 ± 5.7
FA-CA/CS-3	5.5 ± 1.6	7.0 ± 1.9	34.3 ± 7.7

Both the viscosity and surface tension of the solutions sprayed seem to have a combined contribution to the roughness of the final materials. As the viscosity increases, there is a general trend for the roughness to increase (Figure 5a). This result makes sense if the flow of material when reaching the collector is more impeded and the corresponding spreading on the substrate surface is poorer. On the other hand, when surface tension is considered as the variable, the two sets of solutions must be looked at separately to notice the trend (Figure 5b). In both cases, the higher the surface tension, the higher the roughness. The higher the surface tension, the larger the droplets should be, so lower roughness might be expected, at least at the microscale. However, when the surface tension is considered without paying attention to the nature of the liquid, there is no trend, since the liquid with lower viscosity presents higher surface tension. Considering that both surface tension and viscosity interact in a non-linear manner, and their effects can depend on the atomization and evaporation regimes, we can only say that our results point out that viscosity appears to have a greater influence in this specific system, but to confirm the above further analysis—such as rheological measurements during spraying or computational fluid dynamics simulations (CFD simulations) would be necessary. On the other hand, the solvent evaporation rate and polymer–polymer interactions (such as possible formation of segregated chitosan domains) also can influence the morphology. Acetic acid and formic acid differ in volatility and polymer solubility, which can affect the redistribution of cellulose acetate (CA) and chitosan (CS) during droplet drying. Formic acid, being more volatile, promotes faster solvent loss, which can "freeze" local polymer concentrations before thorough mixing occurs, leading to more pronounced microphase segregation. This may result in core–shell or porous structures at the microscale, with CS potentially migrating towards the droplet surface. In contrast, the slower evaporation rate of acetic acid allows polymers to diffuse and rearrange within the droplet, producing a more homogeneous and continuous morphology upon deposition. The interplay between polymer concentration, viscosity, and evaporation kinetics further modulates this behavior, with higher viscosity or polymer loading limiting diffusion and enhancing phase segregation. Understanding these effects is essential for explaining the surface topography, porosity, and microstructural features of films produced by spraying and highlights the importance of solvent selection in controlling the morphology of polymer films prepared by AASS.

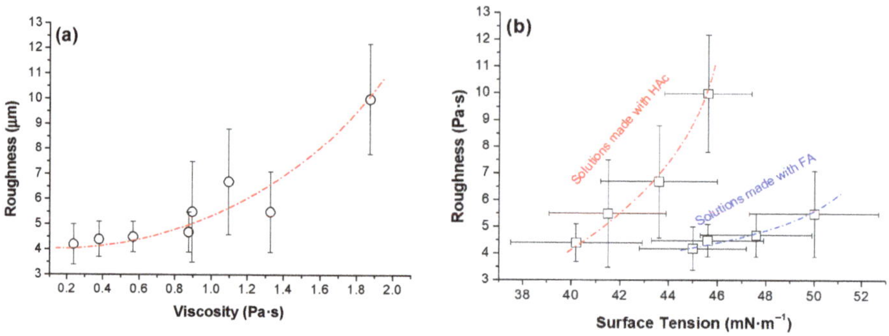

Figure 5. Roughness of AASS materials as a function of solutions' viscosities (**a**) and surface tensions (**b**).

To have higher lateral resolution, FESEM images were taken (Figure 6). It can be observed in all cases that a relatively flat surface whose topography is slightly irregular due to the presence of droplets of material that obtain flat by spreading out when they impact the collector during the spraying process. From the analysis of the microfeatures observed in the FESEM images, it is possible to confirm what was explained above about the contributions of viscosity and surface tension.

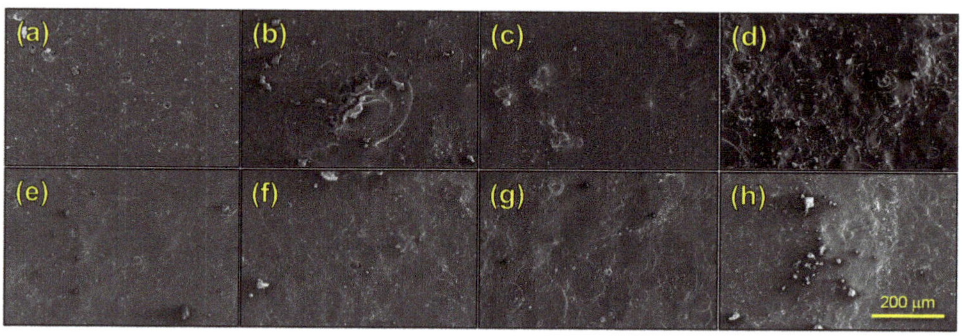

Figure 6. FESEM images of the top side of films produced by AASS, (**a**) HAc-CA; (**b**) HAc-CA/CS-1; (**c**) HAc-CA/CS-2; (**d**) HAc-CA/CS-3; (**e**) FA-CA; (**f**) FA-CA/CS-1; (**g**) FA-CA/CS-2; (**h**) FA-CA/CS-3.

Using ImageJ, the size of droplets observed on the FESEM images was measured. In Figure 7, the distributions of droplet diameters are shown. This information should give a clue with respect to which contribution affects the final roughness of the materials the most. There is a clear tendency to decrease the size of droplets observed in the FESEM images as the viscosity and surface tension of the solution increase, i.e., chitosan concentration increases. For a set of samples made with a particular solvent, the lower the viscosity, the higher the ability for the droplet to spread out, and the lower the surface tension, the larger the droplets can be formed. Therefore, both parameters should contribute in the same direction in terms of the size of microfeatures observed. However, if both sets of samples are compared to each other in terms of materials with the same composition, the viscosity is the factor that conditions the most on the size of the microfeatures observed (Figures 6 and 7). ANOVA test showed a statistically significant difference (with α level 0.05) in the mean values of droplet size between the samples that contain different amounts of chitosan.

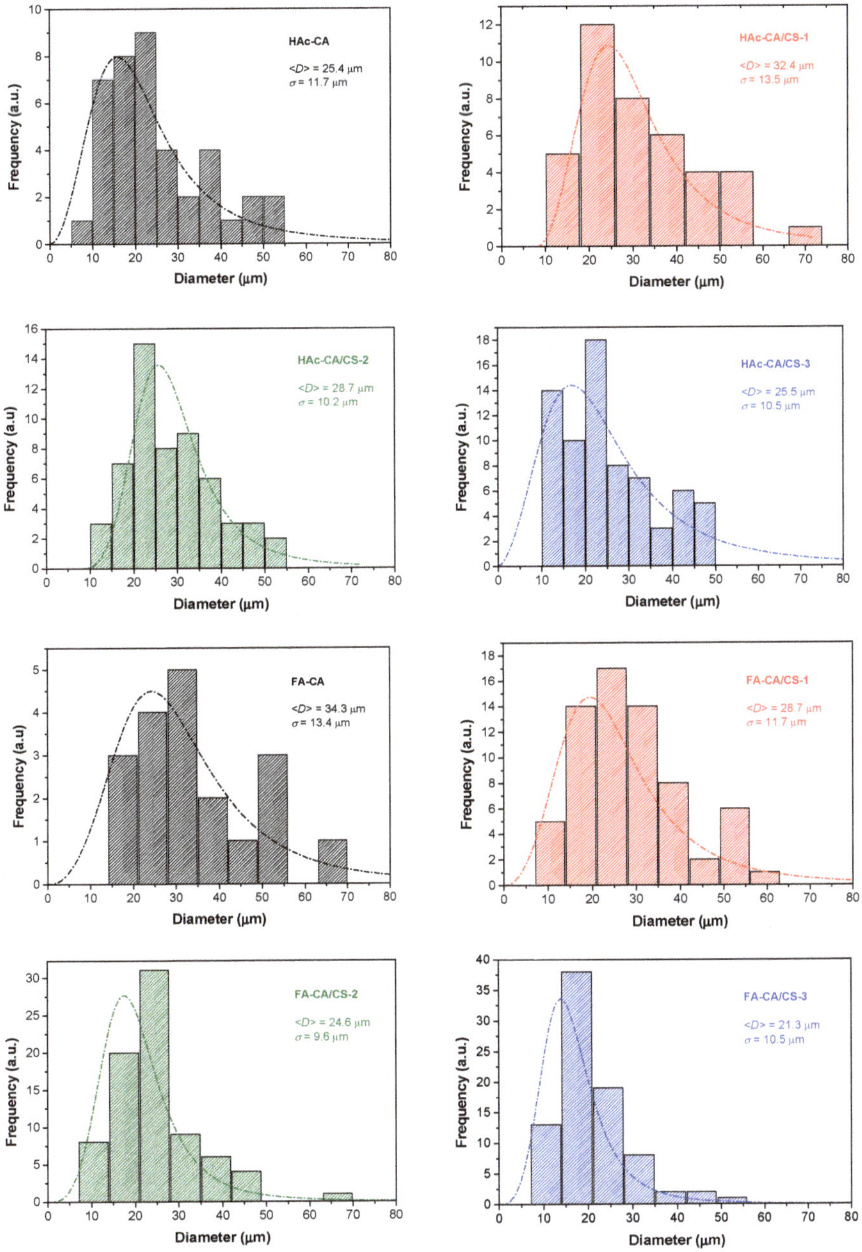

Figure 7. Size distribution of the droplets observed in the FESEM images of the materials analyzed by ImageJ software.

Additionally, to better understand surface properties of the materials, such as solvophilicity behavior, water vapor permeability, and bioactivity, porosity was also analyzed. Figure 8 shows the porosities of all the samples prepared in this study. As can be seen, all materials are quite porous. For a particular composition, there is no clear difference in porosity as a function of the solvent used. However, what is observed is a clear trend to decrease the porosity when the concentration of CS increases, which implies an increase in the surface tension. A possible explanation can be a combination of the viscosity and surface tension increase in the solutions to be sprayed. Although the increase in viscosity must oppose the rapid spreading of the drops on the collector surface, an increase in surface tension must favor a larger droplet size that allows for greater coating efficiency and, consequently, lower porosity. However, it is important here to highlight that the presence of CS may also increase packing density and reduce porosity due to a filling effect or improved matrix cohesion; therefore, porosity reduction could be considered multifactorial, involving increased structural cohesion, reduced spreading ability, and higher packing density of CS-rich droplets.

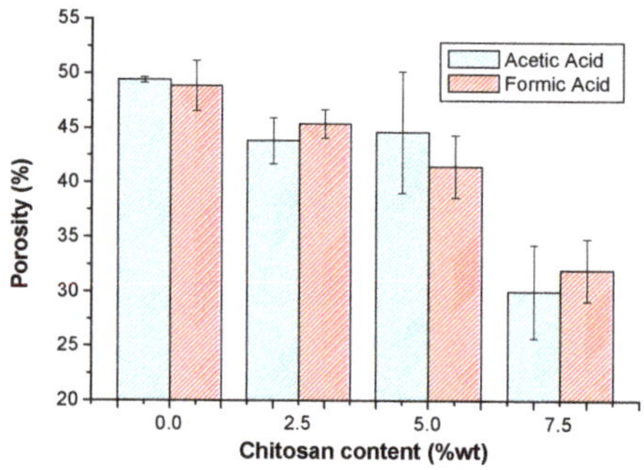

Figure 8. Porosities of all the samples prepared in this study.

3.3. Solvophilicity Behavior

Once the morphology is known in terms of the size and shape of the microfeatures observed together with the porosity, the ability of the materials to be wetted by liquids such as water and oil can be interpreted. Water and oil contact angles were measured, and the corresponding values are represented in Figure 9.

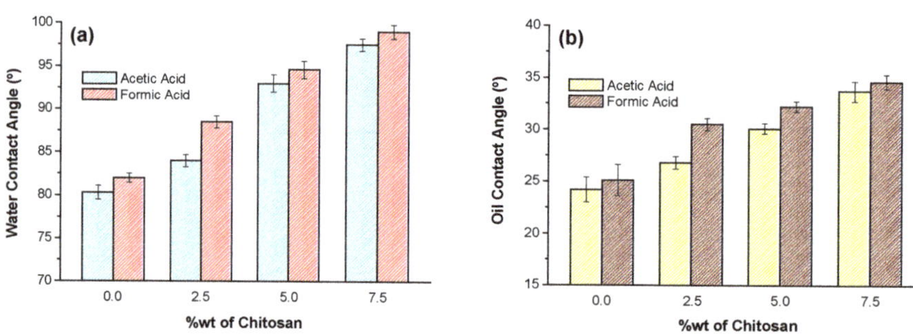

Figure 9. (**a**) Water and (**b**) oil contact angles of droplets on the CA-based materials with different concentrations of chitosan, depending on the solvent used for material preparation, using AASS.

When analyzing contact angles, two contributions must be taken into consideration: physico-chemical interactions and topography. The first contribution should be associated with the nature of the material, and the higher the polarity, the higher the hydrophilicity, which should correspond to a lower water contact angle and higher oil contact angle, respectively. For the materials under consideration, the nature changes as a function of the composition or the concentration of CS. The value of surface tension published for pellets of chitosan with Mw = 58,000 Da and DDA = 67% is 31 $J{\cdot}m^{-2}$ with only dispersive contribution [60] while values of around 50 $J{\cdot}m^{-2}$ were reported for cellulose acetate, where the polar contribution was 17 J·m^{-2} [61]. Therefore, with respect to the surface, the higher the proportion of CA, the higher the hydrophilicity is expected, considering reported data or the simple application of the rule of mixtures [62]. However, the results shown in Figure 9 indicate that both hydrophobicity and oleophobicity increase with higher proportions of CS, which appears to be contradictory. The observed behavior reflects the influence of surface morphology, where increased CS content affects both water and oil contact angles similarly due to changes in porosity and roughness, rather than implying that hydrophilicity and hydrophobicity increase simultaneously in a literal sense. In other words, these results can only be explained through the consideration of morphology as the main contribution to the ability of liquid droplets to spread out on a surface. Then, following the general trends observed, the porosity in combination with roughness should be the parameters that correlate with the contact angles, regardless of the nature of the liquid. What can be seen is that there must be a balance between both parameters since the lower the porosity (Figure 8) and the higher the roughness (Table 3), the higher the contact angle, regardless of the nature of the liquid. However, when comparing materials with the same concentration of CS but prepared with different solvents, an apparent contradiction is observed since, for almost the same porosity (Figure 8), the contact angles are slightly higher (if error is not considered) when the roughness is lower (Figure 9). A possible explanation of this result is to consider the porosity as the parameter that contributes the most to the wettability. The observed contact angles can be explained using the Wenzel [63] and Cassie-Baxter [64] models, which describe how surface morphology influences wettability. While chemical composition suggests that increasing cellulose acetate (CA) should enhance hydrophilicity, our results show that both hydrophilicity and hydrophobicity increase with higher chitosan (CS) content. This can be attributed to the dominant role of surface morphology: porosity controls air entrapment and droplet spreading (Cassie-Baxter effect), while roughness amplifies the intrinsic wettability (Wenzel effect). Thus, the interplay between porosity and roughness, rather than chemical composition alone, primarily determines the observed contact angles.

3.4. Water Vapor Permeability

The water vapor transmission rate is a critical property in food packaging because it prevents moisture transfer, helping to maintain the ideal moisture content inside the packaging. In Figure 10, the values of WVTR obtained for the different materials under study are presented.

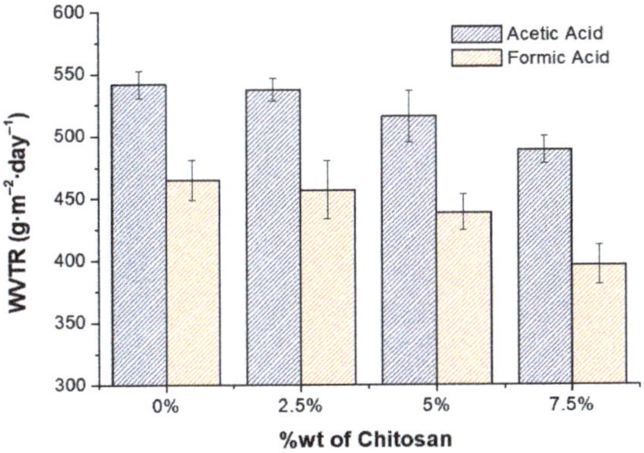

Figure 10. Values of water vapor transmission rate (WVTR) for CA-based materials with varying chitosan concentrations.

As can be seen, the WVTR of all the samples is significantly lower than previously reported for some cellulose acetate-based films [65]. As the concentration of chitosan increases in the composite, the WVTR further decreases. This supports the notion that polar groups in components of composite films can be involved in intra- or intermolecular interactions (e.g., hydrogen bonding) that reduce their availability to interact with water, thus creating a water barrier material. The values of WVTR obtained in this work are significantly lower (550 $g{\cdot}m^{-2}$ day^{-1}), compared with previously reported values for cellulose acetate, and are of the same magnitude as grafted cellulose acetate [66]. In addition, when comparing the samples prepared from solutions made with the different solvents, it is observed that when using FA as the solvent, WVTR is lower, which further supports the hypothesis that lower porosity decreases the ability of the material to transfer water vapor through it.

Jointly observing the results of wettability and water vapor transmission, it can be concluded that the combination of two moderately hydrophilic polymers through processing with solution blowing and creating a specific morphology may lead to obtaining flexible and sustainable materials with properties applicable for food packaging. However, it is important here to recall that moisture sensitivity is a property that should be considered depending on the final application. In certain applications, it is preferred to have higher moisture sensitivity, for example, when using multilayer systems where one of the layers is more moisture sensitive but covered by hydrophobic coatings to regulate vapor transfer. Other examples are cellulose acetate films modified with hydrophilic polymers or salts that absorb moisture, trying to delay mold growth in baked goods. Also, water-soluble or at least partially soluble polymers could be interesting for controlled release systems and edible or biodegradable coatings.

3.5. Mechanical Behavior of Films

Mechanical behavior is another important aspect to control during materials preparation, especially considering the final application. For food packaging, the materials are required to be mechanically consistent but also flexible. Figure 11 shows two representative images of the materials prepared. It is seen that all materials present a high flexibility. Cellulose acetate can be either flexible or brittle depending on several factors, primarily its degree of substitution, the presence of plasticizers, and the processing method. For example, when the degree of substitution is higher than 2.4, as in our case, without plasticizers, films of CA are rigid and brittle. In addition, films obtained by casting or slow drying tend to be more brittle if they do not contain plasticizers. Therefore, the flexible behavior shown by the materials prepared by AASS indicates they are good candidates to be used for food packaging.

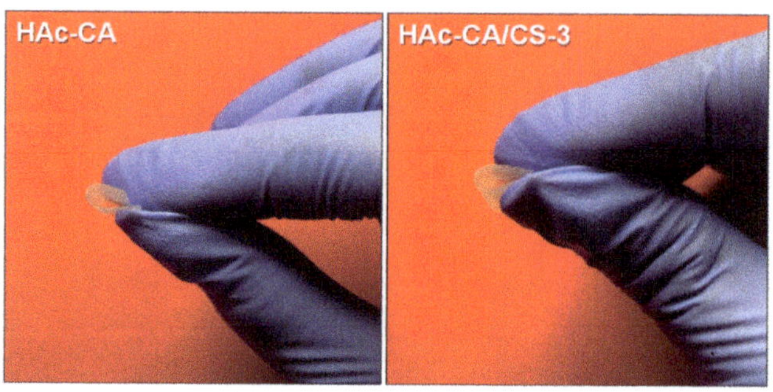

Figure 11. Representative images of the materials prepared, showing their high flexibility.

On the other hand, all materials in this work were tensile tested (Figure S3) to obtain the corresponding strength–strain curves from which values of Young's modulus (E), maximum strength (σ_{max}), and strain at failure (ε_f) were determined. In Table 4, all those mechanical parameters are presented. A one-way ANOVA test showed there is a statistically significant difference (with α level 0.05) in all the mechanical parameters considered in Table 4 when comparing the samples as a function of the concentration of chitosan or when comparing between materials obtained with different solvents.

Table 4. Mechanical parameters obtained from the tensile tests of the CA-based materials prepared by AASS.

Sample	E (Mpa)	σ_{max} (Mpa)	ε_f (%)
Hac-CA	626 ± 14	27 ± 2	4.9 ± 0.5
Hac-CA/CS-1	545 ± 15	43 ± 6	10.3 ± 0.7
Hac-CA/CS-2	482 ± 55	56 ± 3	17.6 ± 1.8
Hac-CA/CS-3	467 ± 21	80 ± 1	19.3 ± 0.8
FA-CA	946 ± 61	19 ± 3	2.1 ± 0.2
FA-CA/CS-1	633 ± 10	37 ± 2	9.9 ± 0.5
FA-CA/CS-2	619 ± 38	44 ± 1	10.2 ± 0.5
FA-CA/CS-3	538 ± 50	64 ± 5	12.9 ± 1.3

As can be seen, both the solvent used to prepare the solutions to be sprayed and the content of chitosan influence the mechanical behavior. Regardless of the solvent used, there is a general trend of increasing the maximum strength and the strain at break of the material by the addition of chitosan, while there is a decrease in the Young's modulus. When acetic acid is used as a solvent, there is an outstanding increase in the maximum strength and strain of the film with the highest chitosan content (7.5%), almost 3 times in the first case and 4 times in the second, compared to the neat cellulose acetate film produced from acetic acid. A similar trend is observed when formic acid is used as the solvent to prepare the materials.

To explain these results, a balance between two contributions can be considered: porosity (Figure 8) and crystallinity degree (Table 2). In principle, the first one should account more for the tensile strength since lower porosity yields more effective cross-sectional area supporting the mechanical load, which in turn should lead to higher tensile strength. On the other hand, the higher the crystallinity degree, the higher the modulus, which may influence the tensile strength. In fact, the highest moduli are found for the samples with the highest crystallinity, which are those prepared by using formic acid as solvent. Despite the above, it is important to highlight the fact that in this work, solution spraying was able to produce flexible and stretchable materials. This behavior may be attributed to the specific microstructure generated by the solution spraying process, which promotes a more homogeneous stress distribution throughout the film and may introduce a fine porous or fibrillar network. Such morphological features reduce brittleness and enable the material to deform without fracturing, thereby enhancing flexibility and stretchability. These properties are crucial for food packaging materials; therefore, the SBS is an excellent choice for flexible film preparation, especially from rigid carbohydrates without the use of any plasticizer.

Here, It Is Important to point out that water content can sometimes Influence the final mechanical behavior under certain conditions; however, the specimens in this work were stored under the same conditions, and no differences were observed in the FTIR spectra in terms of water content (Figure 2). The results obtained in the present study revealed that if all samples have absorbed water, it must be negligible. It is well-known that very small amount of absorbed water in cellulose acetate causes subtle but detectable changes in the IR spectrum such as increase in intensity and broadening of the OH band (3200–3600 cm^{-1}), slight decrease in the ester C=O peak (~1745 cm^{-1}), and minor alterations in the C–O–C band of the acetate (~1230 cm^{-1}). These variations are useful in ATR-FTIR spectroscopy for monitoring moisture changes in cellulose acetate. However, the FTIR spectra obtained for the CA-based materials prepared in the present work do not present those alterations mentioned. As an example, to confirm the latter, Chris Toprak et al. studied the water absorption of cellulose acetate with a degree of acetylation of 2.45 using FTIR in the transmission mode [66]. They found that with a very small amount of water, it is possible to observe clear changes in the absorbance of the bands mentioned above. On the other hand, Murphy et al. [67], using cellulose acetate membranes with the same degree of acetylation and similar molecular weight to the CA used in the present work, reported even higher relative absorbance (obtained by ATR-FITR) of the broad band corresponding to OH stretching for dry samples subjected to a process based on the use of isopropanol to remove the water. As a conclusion, and considering the result obtained from ATR-FTIR spectra in the present work, the water content in the samples under study can be considered negligible, and, as a consequence, no effect should be expected in their mechanical performance.

Finally, it is also important to make a comparison with similar systems obtained by other, more conventional methods, such as solvent casting. For example, A.W. Cindradewi et al. prepared by solvent casting and characterized cellulose acetate films reinforced with cellulose nanofibrils [68]. They obtained for the pure cellulose acetate with a similar degree of acetylation, a tensile strength (80 Mpa), Young's modulus (4 Gpa), and elongation at break (5%), higher than 90%, higher than 80%, and lower 70%, respectively. The main reason for these results must be the porosity, since in the case of solvent casting, it is expected to obtain perfect bulk materials if the rate of evaporation is well controlled. The values of tensile strength and Young's modulus are conditioned by the section area considered to obtain the tensile test curves and, consequently, the mechanical parameters. For porous materials, the real section area that supports the mechanical loads is replaced by the higher section area directly obtained from the measured size of the specimen (width and thickness). However, the important thing here is that enough mechanical consistency is attained by producing the materials by AASS, and that enhanced flexibility is achieved.

3.6. Antimicrobial Activity Against Gram-Negative *E. coli*

Based on previous results presented here, the sample produced from a solution made with formic acid and with 7.5% chitosan can be considered the one with satisfactory properties (mechanical, water barrier, and wettability) to be considered the most appropriate for food packaging applications. Therefore, this sample was selected to study its bioactivity against *E. coli*. Figure 12 shows images of bacterial colonies produced in the corresponding samples processed from formic acid using solution spraying without and with 7.5% of chitosan. The analysis revealed that the number of colonies in the cellulose acetate/chitosan blend is significantly reduced (~2 orders of magnitude lower), indicating a potential antibacterial effect of the material on food preservation. These results testify that the antimicrobial activity of chitosan was retained in the composite matrix, likely due to partial availability of its functional groups, for example, the positively charged amino groups that could interact with the bacterial cell wall [69,70]. This result is also expected based on the fact that the antimicrobial activity of chitosan is frequently reported to be mostly pronounced towards Gram-negative bacteria [71]. This is a consequence of the interaction between positively charged amino groups of chitosan and the negatively charged cell wall of bacteria. If we consider this aspect and accept the possibility of interaction between amino groups of chitosan, we can confirm that in the composite developed in this work, interactions between components are mostly based on OH groups between cellulose acetate and chitosan.

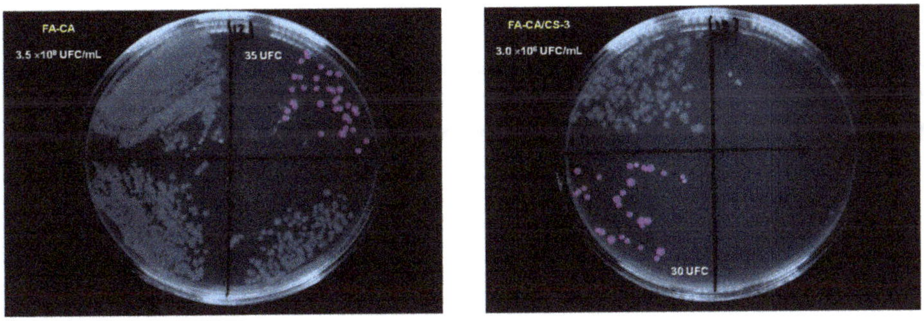

Figure 12. Images of bacterial colonies produced in the samples produced from solutions made with formic acid, without (**left**) and with 7.5% of chitosan (**right**).

4. Conclusions

In this work, the preparation of innovative blends in the form of films from cellulose acetate and chitosan using air-assisted solution spraying (AASS) is presented. Films developed by AASS have a unique morphology consisting of droplets that merge during production, outstanding mechanical properties (up to 80 MPa maximum strength depending on the blend composition and solvent used for materials preparation), high contact angle towards water (up to 95°), and moderate contact angle with oil (<90°). Mentioned properties, taken together with antibacterial activity against E. coli, a Gram-negative bacterium, make these films the perfect candidates for food packaging applications. Considering the increasing need to obtain sustainable materials while decreasing carbon footprint and negative environmental impact, the processing of biopolymers with solution blow spinning represents a significant step toward sustainability goals. The absence of any plasticizer in the formulation while maintaining the flexibility and mechanical integrity of films is a great advantage since the presence of plasticizers is frequently of short duration, and their leaching from the material prevents this kind of material from being used for food packaging. The mechanical and sorption properties, together with bioactivity of developed materials, alongside ease of processing using AASS, testify to the superiority of this technique over others for film preparation (casting, molding) and bring high potential into the sector of food packaging materials. Although the purpose of the present work was not the direct deposition of materials on the food, the information gathered in this study should be potentially useful when understanding the performance of these materials if they are directly deposited by AASS on any target, including food items for their preservation.

Supplementary Materials: The following supporting information can be downloaded at: https://www.mdpi.com/article/10.3390/polym17182479/s1, Table S1: Solubility data for the cellulose acetate (CA), chitosan (CS), and mixtures CA/CS; Figure S1: Representative photos of the CA-based films prepared by solution spraying; Figure S2: X-ray diffraction patterns of the as-received polymers CA and CS, respectively; Figure S3: Stress–strain curves obtained from tensile tests performed for all the specimens of each sample.

Author Contributions: Conceptualization, J.G.-B. and A.K.; methodology, N.M.F., A.K. and J.G.-B.; validation, N.M.F., A.K. and J.G.-B.; formal analysis, N.M.F., A.K. and J.G.-B.; investigation, N.M.F., A.K. and J.G.-B.; resources, A.K. and J.G.-B.; data curation, N.M.F., A.K. and J.G.-B.; writing—original draft preparation, J.G.-B.; writing—review and editing, N.M.F., A.K. and J.G.-B.; visualization, N.M.F., A.K. and J.G.-B.; supervision, A.K. and J.G.-B.; project administration, A.K. and J.G.-B.; funding acquisition, A.K. and J.G.-B. All authors have read and agreed to the published version of the manuscript.

Funding: This research was funded by CONEX-Plus program of Universidad Carlos III de Madrid (UC3M) and the European Commission through the Marie-Sklodowska Curie COFUND Action, grant number 801538, by the Ministerio de Ciencia, Innovación y Universidades of Spain (MICIU/AEI/10.13039/501100011033, Project I + D + i PID2020-112713RB-C22), and by the University Carlos III de Madrid, Fondos de Investigación of Fco. Javier González Benito, grant number 2012/00130/004, and the strategic Action in Multifunctional Nanocomposite Materials, grant number 2011/00287/003.

Data Availability Statement: All data will be available upon request.

Acknowledgments: Bacterial culturing was performed in the laboratories of the Department of Bioengineering. We particularly appreciate the work carried out by the technician Angélica Corral.

Conflicts of Interest: The authors declare no conflicts of interest.

References

1. Shchipunov, Y. Bionanocomposites: Green Sustainable Materials for the near Future. *Pure Appl. Chem.* **2012**, *84*, 2579–2607. [CrossRef]
2. Deng, Y.; Zhu, T.; Cheng, Y.; Zhao, K.; Meng, Z.; Huang, J.; Cai, W.; Lai, Y. Recent Advances in Functional Cellulose-Based Materials: Classification, Properties, and Applications. *Adv. Fiber Mater.* **2024**, *6*, 1343–1368. [CrossRef]
3. Marinho, E. Cellulose: A comprehensive review of its properties and applications. *Sustain. Chem. Environ.* **2025**, *11*, 100283. [CrossRef]
4. Ndlovu, S.P.; Alven, S.; Hlalisa, K.; Aderibigbe, B.A. Cellulose Acetate-Based Wound Dressings Loaded with Bioactive Agents: Potential Scaffolds for Wound Dressing and Skin Regeneration. *Curr. Drug Deliv.* **2024**, *21*, 1226–1240. [CrossRef] [PubMed]
5. Oldal, D.G.; Topuz, F.; Holtzl, T.; Szekely, G. Green Electrospinning of Biodegradable Cellulose Acetate Nanofibrous Membranes with Tunable Porosity. *ACS Sustain. Chem. Eng.* **2023**, *11*, 994–1005. [CrossRef]
6. Zhao, W.; Cao, S.; Cai, H.; Wu, Y.; Pan, Q.; Lin, H.; Fang, J.; He, Y.; Deng, H.; Liu, Z. Chitosan/Silk Fibroin Biomimic Scaffolds Reinforced by Cellulose Acetate Nanofibers for Smooth Muscle Tissue Engineering. *Carbohydr. Polym.* **2022**, *298*, 120056. [CrossRef]
7. Gomes, V.; Pires, A.S.; Mateus, N.; de Freitas, V.; Cruz, L. Pyranoflavylium-Cellulose Acetate Films and the Glycerol Effect towards the Development of PH-Freshness Smart Label for Food Packaging. *Food Hydrocoll.* **2022**, *127*, 107501. [CrossRef]
8. Ravichandran, S.; Sengodan, P.; Saravanan, A.; Vickram, S.; Chopra, H. Antibacterial Food Packaging Using Biocompatible Nickel Oxide-Infused Cellulose Acetate Electrospun Nanofibers. *Food Chem.* **2025**, *472*, 142888. [CrossRef]
9. Puls, J.; Wilson, S.A.; Hölter, D. Degradation of Cellulose Acetate-Based Materials: A Review. *J. Polym. Environ.* **2011**, *19*, 152–165. [CrossRef]
10. Charvet, A.; Vergelati, C.; Long, D.R. Mechanical and Ultimate Properties of Injection Molded Cellulose Acetate/Plasticizer Materials. *Carbohydr. Polym.* **2019**, *204*, 182–189. [CrossRef]
11. Sengupta, T.; Han, J.H. Surface Chemistry of Food, Packaging, and Biopolymer Materials. In *Innovations in Food Packaging*; Elsevier: Amsterdam, The Netherlands, 2014; pp. 51–86.
12. Rafique, A.; Sequeira, I.; Bento, A.S.; Moniz, M.P.; Carmo, J.; Oliveira, E.; Oliveira, J.P.; Marques, A.; Ferreira, I.; Baptista, A.C. A Facile Blow Spinning Technique for Green Cellulose Acetate/Polystyrene Composite Separator for Flexible Energy Storage Devices. *Chem. Eng. J.* **2023**, *464*, 142515. [CrossRef]
13. Bolloli, M.; Antonelli, C.; Molméret, Y.; Alloin, F.; Iojoiu, C.; Sanchez, J.-Y. Nanocomposite Poly(Vynilidene Fluoride)/Nanocrystalline Cellulose Porous Membranes as Separators for Lithium-Ion Batteries. *Electrochim. Acta* **2016**, *214*, 38–48. [CrossRef]
14. Yvonne, T.; Zhang, C.; Zhang, C.; Omollo, E.; Ncube, S. Properties of Electrospun PVDF/PMMA/CA Membrane as Lithium Based Battery Separator. *Cellulose* **2014**, *21*, 2811–2818. [CrossRef]
15. An, Y.; Li, F.; Di, Y.; Zhang, X.; Lu, J.; Wang, L.; Yan, Z.; Wang, W.; Liu, M.; Fei, P. Hydrophobic Modification of Cellulose Acetate and Its Application in the Field of Water Treatment: A Review. *Molecules* **2024**, *29*, 5127. [CrossRef]
16. Ding, B.; Li, C.; Hotta, Y.; Kim, J.; Kuwaki, O.; Shiratori, S. Conversion of an Electrospun Nanofibrous Cellulose Acetate Mat from a Super-Hydrophilic to Super-Hydrophobic Surface. *Nanotechnology* **2006**, *17*, 4332–4339. [CrossRef]
17. Lee, H.; Nishino, M.; Sohn, D.; Lee, J.S.; Kim, I.S. Control of the Morphology of Cellulose Acetate Nanofibers via Electrospinning. *Cellulose* **2018**, *25*, 2829–2837. [CrossRef]
18. Ji, D.; Lin, Y.; Guo, X.; Ramasubramanian, B.; Wang, R.; Radacsi, N.; Jose, R.; Qin, X.; Ramakrishna, S. Electrospinning of Nanofibres. *Nat. Rev. Methods Primers* **2024**, *4*, 1. [CrossRef]

19. Daristotle, J.L.; Behrens, A.M.; Sandler, A.D.; Kofinas, P. A Review of the Fundamental Principles and Applications of Solution Blow Spinning. *ACS Appl. Mater. Interfaces* **2016**, *8*, 34951–34963. [CrossRef]
20. Nikolić, N.; Olmos, D.; González-Benito, J. Key Advances in Solution Blow Spinning of Polylactic-Acid-Based Materials: A Prospective Study on Uses and Future Applications. *Polymers* **2024**, *16*, 3044. [CrossRef]
21. Gao, Y.; Zhang, J.; Su, Y.; Wang, H.; Wang, X.-X.; Huang, L.-P.; Yu, M.; Ramakrishna, S.; Long, Y.-Z. Recent Progress and Challenges in Solution Blow Spinning. *Mater. Horiz.* **2021**, *8*, 426–446. [CrossRef]
22. Kramar, A.; Luxbacher, T.; González-Benito, J. Solution Blow Co-Spinning of Cellulose Acetate with Poly(Ethylene Oxide). Structure, Morphology, and Properties of Nanofibers. *Carbohydr. Polym.* **2023**, *320*, 121225. [CrossRef]
23. Kramar, A.; González-Benito, J. Preparation of Cellulose Acetate Film with Dual Hydrophobic-Hydrophilic Properties Using Solution Blow Spinning. *Mater. Des.* **2023**, *227*, 111788. [CrossRef]
24. Claro, P.I.C.; Cunha, I.; Paschoalin, R.T.; Gaspar, D.; Miranda, K.; Oliveira, O.N.; Martins, R.; Pereira, L.; Marconcini, J.M.; Fortunato, E.; et al. Ionic Conductive Cellulose Mats by Solution Blow Spinning as Substrate and a Dielectric Interstrate Layer for Flexible Electronics. *ACS Appl. Mater. Interfaces* **2021**, *13*, 26237–26246. [CrossRef] [PubMed]
25. Dadol, G.C.; Kilic, A.; Tijing, L.D.; Lim, K.J.A.; Cabatingan, L.K.; Tan, N.P.B.; Stojanovska, E.; Polat, Y. Solution Blow Spinning (SBS) and SBS-Spun Nanofibers: Materials, Methods, and Applications. *Mater. Today Commun.* **2020**, *25*, 101656. [CrossRef]
26. Gharsallaoui, A.; Joly, C.; Oulahal, N.; Degraeve, P. Nisin as a Food Preservative: Part 2: Antimicrobial Polymer Materials Containing Nisin. *Crit. Rev. Food Sci. Nutr.* **2016**, *56*, 1275–1289. [CrossRef]
27. Kumar, S.; Mukherjee, A.; Dutta, J. Chitosan Based Nanocomposite Films and Coatings: Emerging Antimicrobial Food Packaging Alternatives. *Trends Food Sci. Technol.* **2020**, *97*, 196–209. [CrossRef]
28. Atarés, L.; Chiralt, A. Essential Oils as Additives in Biodegradable Films and Coatings for Active Food Packaging. *Trends Food Sci. Technol.* **2016**, *48*, 51–62. [CrossRef]
29. Cao, G.; Lin, H.; Kannan, P.; Wang, C.; Zhong, Y.; Huang, Y.; Guo, Z. Enhanced Antibacterial and Food Simulant Activities of Silver Nanoparticles/Polypropylene Nanocomposite Films. *Langmuir* **2018**, *34*, 14537–14545. [CrossRef]
30. Zhang, W.; Sani, M.A.; Zhang, Z.; McClements, D.J.; Jafari, S.M. High Performance Biopolymeric Packaging Films Containing Zinc Oxide Nanoparticles for Fresh Food Preservation: A Review. *Int. J. Biol. Macromol.* **2023**, *230*, 123188. [CrossRef]
31. Zhang, W.; Rhim, J.-W. Titanium Dioxide (TiO2) for the Manufacture of Multifunctional Active Food Packaging Films. *Food Packag. Shelf Life* **2022**, *31*, 100806. [CrossRef]
32. Verlee, A.; Mincke, S.; Stevens, C.V. Recent Developments in Antibacterial and Antifungal Chitosan and Its Derivatives. *Carbohydr. Polym.* **2017**, *164*, 268–283. [CrossRef] [PubMed]
33. Sharma, N.; Modak, C.; Singh, P.K.; Kumar, R.; Khatri, D.; Singh, S.B. Underscoring the Immense Potential of Chitosan in Fighting a Wide Spectrum of Viruses: A Plausible Molecule against SARS-CoV-2? *Int. J. Biol. Macromol.* **2021**, *179*, 33–44. [CrossRef] [PubMed]
34. Zhou, H.; Tong, H.; Lu, J.; Cheng, Y.; Qian, F.; Tao, Y.; Wang, H. Preparation of bio-based cellulose acetate/chitosan composite film with oxygen and water resistant properties. *Carbohydr. Polym.* **2021**, *270*, 118381. [CrossRef] [PubMed]
35. Cazón, P.; Vázquez, M. Mechanical and Barrier Properties of Chitosan Combined with Other Components as Food Packaging Film. *Environ. Chem. Lett.* **2020**, *18*, 257–267. [CrossRef]
36. Kurek, M.; Ščetar, M.; Voilley, A.; Galić, K.; Debeaufort, F. Barrier Properties of Chitosan Coated Polyethylene. *J. Memb. Sci.* **2012**, *403–404*, 162–168. [CrossRef]
37. Leceta, I.; Guerrero, P.; de la Caba, K. Functional Properties of Chitosan-Based Films. *Carbohydr. Polym.* **2013**, *93*, 339–346. [CrossRef]
38. Samios, E.; Dart, R.K.; Dawkins, J.V. Preparation, Characterization and Biodegradation Studies on Cellulose Acetates with Varying Degrees of Substitution. *Polymer* **1997**, *38*, 3045–3054. [CrossRef]
39. Yuan, Y.; Chesnutt, B.M.; Haggard, W.O.; Bumgardner, J.D. Deacetylation of Chitosan: Material Characterization and in Vitro Evaluation via Albumin Adsorption and Pre-Osteoblastic Cell Cultures. *Materials* **2011**, *4*, 1399–1416. [CrossRef]
40. Domínguez, J.E.; Olivos, E.; Vázquez, C.; Rivera, J.M.; Hernández-Cortes, R.; González-Benito, J. Automated Low-Cost Device to Produce Sub-Micrometric Polymer Fibers Based on Blow Spun Method. *HardwareX* **2021**, *10*, e00218. [CrossRef]
41. *ISO 4288:1996*; Geometrical Product Specifications (GPS), Surface Texture: Profile Method, Rules and Procedures for the Assessment of Surface Texture. ISO (International Organization for Standardization): Geneva, Switzerland, 1996.
42. Nikolić, N.; Olmos, D.; Kramar, A.; González-Benito, J. Effect of Collector Rotational Speed on the Morphology and Structure of Solution Blow Spun Polylactic Acid (PLA). *Polymers* **2024**, *16*, 191. [CrossRef]
43. Wypych, G. *Handbook of Polymers*, 2nd ed.; Elsevier: Amsterdam, The Netherlands, 2016; ISBN 9781895198928.
44. Lupa, D.; Płaziński, W.; Michna, A.; Wasilewska, M.; Pomastowski, P.; Gołębiowski, A.; Buszewski, B.; Adamczyk, Z. Chitosan Characteristics in Electrolyte Solutions: Combined Molecular Dynamics Modeling and Slender Body Hydrodynamics. *Carbohydr. Polym.* **2022**, *292*, 119676. [CrossRef]

45. *ISO 2528:2017*; Sheet Materials, Determination of Water Vapour Transmission Rate (WVTR), Gravimetric (Dish) Method. ISO (International Organization for Standardization): Geneva, Switzerland, 2017.
46. Cerqueira, D.A.; Rodrigues Filho, G.; Assunção, R.M.N. A New Value for the Heat of Fusion of a Perfect Crystal of Cellulose Acetate. *Polym. Bull.* **2006**, *56*, 475–484. [CrossRef]
47. *ISO 22196:2011*; Measurement of Antibacterial Activity on Plastics and Other Non-Porous Surfaces. ISO (International Organization for Standardization): Geneva, Switzerland, 2011.
48. *ASTM E2149-20*; Standard Test Method for Determining the Antimicrobial Activity of Antimicrobial Agents Under Dynamic Contact Conditions. ASTM International: West Conshohocken, PA, USA, 2020.
49. Montoya-Escobar, N.; Ospina-Acero, D.; Velásquez-Cock, J.A.; Gómez-Hoyos, C.; Serpa Guerra, A.; Gañan Rojo, P.F.; Vélez Acosta, L.M.; Escobar, J.P.; Correa-Hincapié, N.; Triana-Chávez, O.; et al. Use of Fourier Series in X-Ray Diffraction (XRD) Analysis and Fourier-Transform Infrared Spectroscopy (FTIR) for Estimation of Crystallinity in Cellulose from Different Sources. *Polymers* **2022**, *14*, 5199. [CrossRef]
50. Prakash, J.; Venkataprasanna, K.S.; Bharath, G.; Banat, F.; Niranjan, R.; Venkatasubbu, G.D. In-Vitro Evaluation of Electrospun Cellulose Acetate Nanofiber Containing Graphene Oxide/TiO2/Curcumin for Wound Healing Application. *Colloids Surf. A Physicochem. Eng. Asp.* **2021**, *627*, 127166. [CrossRef]
51. Zhang, Y.; Xue, C.; Xue, Y.; Gao, R.; Zhang, X. Determination of the Degree of Deacetylation of Chitin and Chitosan by X-Ray Powder Diffraction. *Carbohydr. Res.* **2005**, *340*, 1914–1917. [CrossRef]
52. Fei, P.; Liao, L.; Cheng, B.; Song, J. Quantitative Analysis of Cellulose Acetate with a High Degree of Substitution by FTIR and Its Application. *Anal. Methods* **2017**, *9*, 6194–6201. [CrossRef]
53. Scandola, M.; Ceccorulli, G. Viscoelastic Properties of Cellulose Derivatives: 1. Cellulose Acetate. *Polymer* **1985**, *26*, 1953–1957. [CrossRef]
54. Zafar, M.; Ali, M.; Khan, S.M.; Jamil, T.; Butt, M.T.Z. Effect of Additives on the Properties and Performance of Cellulose Acetate Derivative Membranes in the Separation of Isopropanol/Water Mixtures. *Desalination* **2012**, *285*, 359–365. [CrossRef]
55. de Freitas, R.R.M.; Senna, A.M.; Botaro, V.R. Influence of Degree of Substitution on Thermal Dynamic Mechanical and Physicochemical Properties of Cellulose Acetate. *Ind. Crops Prod.* **2017**, *109*, 452–458. [CrossRef]
56. Erdmann, R.; Kabasci, S.; Heim, H.-P. Thermal Properties of Plasticized Cellulose Acetate and Its β-Relaxation Phenomenon. *Polymers* **2021**, *13*, 1356. [CrossRef]
57. Mucha, M.; Pawlak, A. Thermal Analysis of Chitosan and Its Blends. *Thermochim. Acta* **2005**, *427*, 69–76. [CrossRef]
58. Jiang, L.; Huang, X.; Tian, C.; Zhong, Y.; Yan, M.; Miao, C.; Wu, T.; Zhou, X. Preparation and Characterization of Porous Cellulose Acetate Nanofiber Hydrogels. *Gels* **2023**, *9*, 484. [CrossRef]
59. Yamashita, Y.; Endo, T. Deacetylation Behavior of Binary Blend Films of Cellulose Acetate and Various Polymers. *J. Appl. Polym. Sci.* **2006**, *100*, 1816–1823. [CrossRef]
60. Cunha, A.G.; Fernandes, S.C.M.; Freire, C.S.R.; Silvestre, A.J.D.; Neto, C.P.; Gandini, A. What Is the Real Value of Chitosan's Surface Energy? *Biomacromolecules* **2008**, *9*, 610–614. [CrossRef] [PubMed]
61. Gupta, B.S.; Whang, H.S. Surface Wetting and Energy Properties of Cellulose Acetate, Polyester and Polypropylene Fibers. *Int. Nonwovens J.* **1999**, *1*, os-8. [CrossRef]
62. Liu, C.; Bai, R. Preparation of Chitosan/Cellulose Acetate Blend Hollow Fibers for Adsorptive Performance. *J. Memb. Sci.* **2005**, *267*, 68–77. [CrossRef]
63. Wenzel, R.N. Resistance of solid surfaces to wetting by water. *Ind. Eng. Chem.* **1936**, *28*, 988–994. [CrossRef]
64. Cassie, A.B.D.; Baxter, S. Wettability of porous surfaces. *Trans. Faraday Soc.* **1944**, *40*, 546–555. [CrossRef]
65. Tedeschi, G.; Guzman-Puyol, S.; Paul, U.C.; Barthel, M.J.; Goldoni, L.; Caputo, G.; Ceseracciu, L.; Athanassiou, A.; Heredia-Guerrero, J.A. Thermoplastic Cellulose Acetate Oleate Films with High Barrier Properties and Ductile Behaviour. *Chem. Eng. J.* **2018**, *348*, 840–849. [CrossRef]
66. Toprak, C.; Agar, J.N.; Falk, M. State of Water in Cellulose Acetate Membranes. *J. Chem. Soc. Faraday Trans. 1 Phys. Chem. Condens. Phases* **1979**, *75*, 803. [CrossRef]
67. Murphy, D.; de Pinho, M.N. An ATR-FTIR Study of Water in Cellulose Acetate Membranes Prepared by Phase Inversion. *J. Memb. Sci.* **1995**, *106*, 245–257. [CrossRef]
68. Cindradewi, A.W.; Bandi, R.; Park, C.-W.; Park, J.-S.; Lee, E.-A.; Kim, J.-K.; Kwon, G.-J.; Han, S.-Y.; Lee, S.-H. Preparation and Characterization of Cellulose Acetate Film Reinforced with Cellulose Nanofibril. *Polymers* **2021**, *13*, 2990. [CrossRef]
69. Muthu, M.; Pushparaj, S.S.C.; Gopal, J.; Sivanesan, I. A Review on the Antimicrobial Activity of Chitosan Microspheres: Milestones Achieved and Miles to Go. *J. Mar. Sci. Eng.* **2023**, *11*, 1480. [CrossRef]

70. Guarnieri, A.; Triunfo, M.; Scieuzo, C.; Ianniciello, D.; Tafi, E.; Hahn, T.; Zibek, S.; Salvia, R.; De Bonis, A.; Falabella, P. Antimicrobial Properties of Chitosan from Different Developmental Stages of the Bioconverter Insect *Hermetia illucens*. *Sci. Rep.* **2022**, *12*, 8084. [CrossRef]
71. Ke, C.-L.; Deng, F.-S.; Chuang, C.-Y.; Lin, C.-H. Antimicrobial Actions and Applications of Chitosan. *Polymers* **2021**, *13*, 904. [CrossRef]

polymers

MDPI

Article

Enhanced Mechanical Robustness of Sprayed Cellulose Nanofibril Coatings Through Internal Crosslinking with Boric Acid

Pieter Samyn [1,*], Patrick Cosemans [1], Erik V. Van der Eycken [2,3] and Guglielmo A. Coppola [2]

1 Department of Innovations in Circular Economy and Renewable Materials, SIRRIS, Gaston Geenslaan 8, B-3001 Leuven, Belgium; patrick.cosemans@sirris.be

2 Laboratory for Organic & Microwave-Assisted Chemistry (LOMAC), Department of Chemistry, University of Leuven (KU Leuven), Celestijnenlaan 200F, B-3001 Leuven, Belgium

3 Department of Organic Chemistry, Peoples' Friendship University of Russia (RUDN University), Miklukho-Maklaya Street 6, 117198 Moscow, Russia

* Correspondence: pieter.samyn@sirris.be

Abstract

The enhanced mechanical durability of sprayed nanocellulose coatings at the macroscopic level is primarily required to promote their application in demanding industrial applications with frequently exposed surfaces. In this study, different coating configurations are designed by spraying aqueous cellulose nanofiber (CNF) suspensions in combination with boric acid (BA) as an internal crosslinker and polydopamine (PDA) as an adhesive interlayer onto glass substrates. Multilayer coatings (CNF/BA) or mixed-layer coatings (CNF + BA) with various concentrations of BA and numbers of sprayed layers are evaluated for maximized mechanical performance based on tape tests, rub tests, cross-cut tests, and scratching tests. Good adhesive strength was realized with an interlayer of PDA/BA (high-concentration BA = 10 mM). The highest cohesive strength was observed for a mixed CNF + BA coating (high-concentration BA = 10 mM) with a scratch resistance of 9 N, and a multilayer CNF/BA coating (gradient layers with ultra-high BA concentration = 100 mM) with a scratch resistance of 8 N. The coatings with the highest density did not uniquely introduce the best mechanical resistance when comparing CNF/BA and CNF + BA coatings, as the formation of BA crystals in multilayer coatings might negatively affect the mechanical properties through embrittlement. Alternatively, the mixed CNF + BA coatings with high BA concentrations provide high density and the best mechanical resistance. The favorable crosslinking corresponds to stabilized water contact angles and reduced spreading of the water as a function of time, while a decrease in coating density causes a reduction in transparency. The chemical interactions between CNF and BA are illustrated by infrared spectroscopy, confirming a reduction in free hydroxyl groups upon crosslinking.

Keywords: nanocellulose; coating; crosslinking; boric acid; mechanical testing

1. Introduction

With the increasing availability of nanocellulose materials at an industrial scale, these materials have become favorable candidates for common use as protective bio-based coatings. Aqueous suspensions of cellulose nanofibrils (CNFs) with appropriate viscosity and shear thinning properties allow for the homogeneous deposition of sprayed coatings on large surfaces [1]. Nowadays, nanocellulose coatings are primarily applied to create effective gas barriers against oxygen and water vapor in paper, board, and textile applications [2,3]. Alternatively, the nanoporous coating structure can be used as a host-carrier material for local integration of active species such as drugs, absorbents, or antimicrobial ingredients [4]. For more demanding conditions, however, challenges remain in optimizing their long-term performance when exposed to light, moisture, or mechanical stress [5]. In particular, mechanical robustness and durability need to be improved to enhance the coating lifetime and integrity when they are used in, e.g., manufacturing, construction, or high-touch surfaces [4]. As the coatings of native CNFs are initially formed through random organization of long fibrils, they provide intrinsic mechanical resistance depending on the mechanical entanglements and physical interactions (i.e., hydrogen bonding) between the individual fibrils. Despite the high mechanical strength of single fibrils [6], and the theoretical strength of hydrogen bonds in nanocellulose assemblies [7], however, the macroscale mechanical resistance of the sprayed coatings rather depends on the internal structural morphology developed after application [8]. Moreover, the compatibility of a nanocellulose coating with inert or mineral construction materials becomes more challenging and often requires surface grafting of silanes [9], as opposed to deposition on porous fiber-like substrates [10,11]. In particular, several methods to improve coating adhesion and stability on polymer films and textiles using physical and chemical crosslinking with polymers and polycarboxylic acids have been analyzed [12]. However, the low density and high porosity of textiles and paper are beneficial in terms of strong adhesion of the coatings as compared to flat inert substrates. Therefore, the improvement in both cohesive and adhesive properties of nanocellulose coatings needs to be addressed for high mechanical durability and robustness in daily use.

The pathways to improve mechanical, thermal, and structural stability of nanocellulose coatings by crosslinking include non-covalent (physical) or covalent (chemical) binding [13]. The mechanisms for physical crosslinking are based on charge interactions and/or hydrogen bonding, which require the additional surface modification of the nanocellulose, e.g., through the introduction of anionic carboxyl groups during TEMPO oxidation [14]. The latter often provides relatively weak properties or swelling [15]. The crosslinking of anionic nanocellulose (e.g., after oxidation) with cationic ions (e.g., H^+ and Al^{3+}) resulted in a paper coating with antimicrobial properties and reduced water absorption [16]. In composite membranes, the direct crosslinking between charged nanocellulose fibers with carboxylic groups and deacetylate chitin nanofibers resulted in better mechanical strength [17]. The reaction between the carboxylic acid and amine groups introduced during microwave irradiation also improved strength and hydrolysis resistance [18]. The crosslinking of carboxyl groups in TEMPO-CNF with polyethyleneimine (PEI) or hexamethylenediamine (HMDA) could enhance stiffness and reduce the swelling capacity of CNF foams [15]. Alternatively, crosslinking through direct interaction with the abundantly available hydroxyl groups of native nanocellulose would reduce the need for surface modification by either oxidation or grafting. Crosslinking with guar gum as a dialdehyde occurred through interaction with one or two hydroxyl groups of the nanocellulose providing mechanical reinforcement for packaging films [19], but was not transferrable to coatings due to the strong increase in viscosity under operational conditions and weak robustness in combination with the native (non-oxidized) nanocellulose fibrils [20]. Chemical crosslinking through esterification is a widely employed strategy for the modification of nanocellulose surfaces while preserving their structural integrity. This technique utilizes various reagents such as toluene sulfonic acid and hexanoic acid [21], or possible acylating agents such as butyryl, benzoyl, naphthoyl, diphenylacetyl, and stearoyl groups [22]. In recent developments, both citric acid and succinic acid were used as poly(carboxylic acid) crosslinkers for nanocellulose films [23], due to their nontoxic profile and environmental compatibility. However, the crosslinking with carboxylic acids mainly enhanced the wet strength of cellulose-based substrates [24,25], and reduced the aqueous solubility [26], while the crosslinked nanocellulose papers turned brittle after drying [25]. The most stable nanocellulose films were formed by citric acid crosslinking at high acid concentrations (>20%) in combination with heating above 120 °C, as the crosslinker simultaneously acted as a plasticizer in films with increased water vapor sorption and reduced tensile strength when no thermal heating was applied [27]. The introduction of acid crosslinkers (e.g., butane tetracarboxylic acid) as a source of anionic surface groups in nanocellulose membranes could simultaneously provide additional intrinsic properties such as conductivity, ion transport, and/or selective absorption depending on the crosslinking density [28]. After all, the chemistry of the nanocellulose coatings needs optimization during formulation and application as a higher crosslinking density may result in deteriorated mechanical properties and brittleness [29], lower biodegradability [30], or higher viscosity [31], or occupation of hydroxyl groups by crosslinking may reduce access for additional functionalization [32]. Therefore, a good balance in the degree of crosslinking should be obtained [33], or coating toughness should be altered with plasticizers such as sorbitol [34] or polyethylene glycol [35].

As an alternative environmentally benign crosslinker, boric acid (BA) is considered as a common form of boron that is naturally present in various minerals and which is nontoxic to organisms at appropriately low concentrations [36]. The favorable chemical interactions between boronic acids (a boron atom with two hydroxyl groups linked to an organic molecule) and nanocellulose were investigated, where the boron center forms covalent bonds with the hydroxylated polymer under mild conditions [37]. These bonds are reversible and influenced by the surrounding pH, allowing for a variety of applications [38]. Nonetheless, chemical modification of the polymer or molecule of interest is required to establish the boronic acid moiety. Boric acid (one boron atom linked to three hydroxyl groups) can react subsequently with two diols to form a tetrahedral complex [39], which may suggest enhanced reactivity of the latter [40]. This happens through similar mechanisms as for boronic acid and allows crosslinking within cellulose chains [41]. The interactions of borate anions with vicinal diols in the glucopyranoside moieties of cellulose could be tailored to yield chemisorbed boron chelate complexes [42], but the local cellulose backbone and side chain flexibility as well as the structures of polyol groups play a significant role in determining the kinetic and thermodynamic stability of chelate complexes [43]. Therefore, the interaction may depend on the specific morphology of the nanocellulose and its structural configuration (e.g., coatings, membranes, foams, films, or hydrogels). BA has particularly been used in crosslinking of microfibrillated cellulose composites [44], reinforcement of (nano-)cellulose particleboards [45–47], flame-retardant nanocellulose foams [48], nanocellulose hydrogels [49], antibacterial CNF hydrogels for drug loading [50], self-healing CNF hydrogels [51], nanocellulose adhesives [52], and food packaging films of cellulose nanocrystals and polyvinyl alcohol with antimicrobial properties [53]. The crosslinked films of hybrid CNFs and BA indeed introduced enhanced thermo-mechanical stability with additional flame retardance [54], or potentially antibacterial properties [55]. Furthermore, the addition of BA in various concentrations to nanocellulose composite coatings increased the surface hardness of coated wooden surfaces while the water contact angles declined [56]. However, the governing parameters in crosslinking of sprayed nanocellulose coatings with BA have not yet been further investigated and optimized. In combination with this, polydopamine (PDA) serves as a universal bio-inspired adhesive [57], acting as a chemical compatibilizer due to the presence of catechol groups forming complexes and chemical bonds [58]. The versatility and modulation of a self-polymerized PDA layer acting as a reactive species towards inert substrates such as glass [59] or nanocellulose [60] could substantiate its role as an adhesive interlayer. Although PDA is supposed to simultaneously serve as a crosslinker in nanocellulose hydrogels [61], its reactivity strongly depends on the oxidation state and environmental conditions (e.g., humidity, wet/dry, pH), being more difficult to control as compared to boric acid salts with a higher reactivity [62].

In this paper, the mechanical robustness of sprayed nanocellulose coatings onto glass was improved through the design of different coating configurations with either multilayers or mixed layers, in combination with boric acid (BA) as internal crosslinker and polydopamine (PDA) as adhesive interlayer. While acid crosslinkers for nanocellulose materials have been identified before, reference data for their mechanical performance as coatings is scarce. After determination of appropriate coating systems with maximized mechanical performance, the physical properties and chemical interactions between nanocellulose and BA crosslinkers at appropriate concentrations are further illustrated. Whereas most available studies relate to durability of nanocellulose coatings in textile applications, this study aims at progressing the performance of sprayed nanocellulose coatings under demanding conditions in manufacturing and construction.

2. Materials and Methods

2.1. Materials and Chemicals

The substrates to be coated were microscopic borosilicate glass slides with sizes of 25×75 mm^2 (VWR Belgium, Leuven, Belgium). Dopamine (DA) hydrochloride was obtained from TCI Europe NV (Zwijndrecht, Belgium) and dissolved at a concentration of 2 mg/mL in 10 mM Tris–HCl buffer at pH 8.5 (Acros Organics, Geel, Belgium). The glass substrates were first coated with a polydopamine (PDA) layer by dipping in the DA solution for 24 h. Alternatively, the PDA coating was deposited through spraying with a solution of dopamine hydrochloride and sodium periodate (Acros Organics, Geel, Belgium). The glass/PDA substrates were thoroughly rinsed with ultra-pure water (milliQ, 18.2 µs/cm, Merck, Darmstadt, Germany) and dried at room temperature. More information on deposition of a PDA interlayer onto glass and its characterization can be found elsewhere [63,64].

The boric acid (BA) was obtained from Merck Life Science BV (Hoeilaart, Belgium) and prepared in stock solutions with three different concentrations of 1 mM (low BA), 10 mM (high BA), and 100 mM (ultra-high BA), and with the addition of one equivalent of sodium hydroxide (Merck Life Science BV, Hoeilaart, Belgium). The glass/PDA substrates were additionally coated by dipping or spraying the substrate with a low-BA or high-BA solution, followed by air drying of the glass/PDA/BA substrates.

The cellulose nanofibrils (CNFs) of Valida® fine grade S191C (Sappi, Maastricht, The Netherlands) were produced as patented under industrial conditions of mechanical fibrillation at pilot scale [65], and delivered as an aqueous suspension with dry solid content of 3 wt.% (see Supplementary Figure S1).

2.2. Methods

The nanocellulose coating systems were deposited onto glass/PDA/BA substrates and fabricated in two configurations, of (i) multilayer coatings (CNF/BA) with alternating sprayed layers of CNF and BA crosslinker, or (ii) mixed-layer coatings (CNF + BA) with sprayed layers of a prepared CNF + BA suspension. The multilayer coatings were prepared by diluting the 3 wt.% CNF suspension to 1.5 wt.% CNF by adding demineralized water. The mixed-layer coatings were prepared by diluting the 3 wt.% CNF suspension with either the low-BA, high-BA, or ultra-high-BA solution, respectively, in a 1:1 weight ratio. The suspensions were mixed with a high-shear mixer (Orange-line Dispermill, ATP Engineering, Almere, The Netherlands) at 3000 rpm for 30 min and remained stable during spraying under selected conditions.

The coating layers were deposited with a manual air spray gun, 1000 RP (SATA GmbH, Kornwestheim, Germany), using a 1.6 mm nozzle diameter and 1.5 bar dynamic pressure after preliminary optimization of the spraying parameters. The spraying was performed in a spraying booth with controlled airflow and a temperature of 25 °C. The glass substrates were mounted on a horizontal table with perpendicular orientation of the spray gun at 20 cm above the samples, while applying a horizontally advancing gun speed of 10 mm/s. The multilayer coatings and mixed-layer coatings were sprayed in thicknesses corresponding to 3, 5, and 10 layers of CNF with alternating BA layers (multilayers), or 3, 5, and 10 sprayed layers of CNF with mixed BA (mixed layers). The respective layers were deposited in "wet" conditions (no intermediate drying in between the sprayed layers, final drying for 30 min at 70 °C after spraying the final CNF top layer) or "dry" conditions (with intermediate drying after each deposited layer for 30 min at 70 °C in a circulating hot-air oven). The samples were stored under a conditioned atmosphere (23 °C, 50% RH) before further characterization.

2.3. Characterization Techniques

The mechanical resistance of the coatings was evaluated by a micro-scratching test using the sclerometer type 3092 (Elcometer Instruments GmbH, Aalen, Germany) with a tungsten carbide tip of 0.75 mm radius loaded under 0 to 11 N by exchange of the spring with an appropriate spring constant, following ISO 1518-1 [66]. The stylus travelled at a speed of 35 mm/s over a length of the scratching track of 40 mm. The cross-cut testing was applied according to ISO 2409:2020 [67], using the cross-hatch adhesion tester type 1542 (Elcometer Instruments GmbH, Aalen, Germany) with 2 mm cutter spacing and adhesive tape. The performance results are quantified by scoring grades 0 (good adherence) to 5 (failure). The tape test was performed on either a native coating or after cross-cut testing by peeling in a 90° perpendicular direction. A fingernail rubbing test was performed by scratching one stroke in agreement with the DBL 7383 specifications. Microscopic evaluation of the mechanical damage was performed through optical microscopy on a Keyence VHX-7000 (Keyence, Mechelen, Belgium) with ×20 to ×200 magnification of objective lenses.

The detailed morphologies and 3D topographies were studied on a confocal laser scanning microscope VK-X3000 (Keyence, Mechelen, Belgium) with ×10 to ×150 objective lenses (additionally, with up to ×8 digital magnification). The surface roughness parameters (Sa = average surface roughness) were determined on images at ×20 or ×50 magnification, after a consistent surface flattening procedure eliminating the long-range waviness, following ISO 25178 [68]. The coating thickness was determined as an average of 10 height differences measured from line scans over a cross-sectional scratch at ×20 magnification. The weight of the different coating layers was measured on an analytical balance (Sartorius, Göttingen, Germany) with 0.001 mg precision.

The static water contact angles were determined with an OCA50 goniometer (Dataphysics Instruments GmbH, Filderstadt, Germany) after depositing a 3 µL droplet of deionized water and shape fitting with a tangent fit procedure for hydrophilic surfaces according to ISO 19403-2 [69]. The evolution of the contact angle during spreading of the water droplet over the surface was followed for 20 s. UV/VIS spectroscopy was performed on a UV-2450 spectrophotometer (Shimadzu, Kyoto, Japan) in transmission mode recording the transmission spectra over the 200 to 900 nm wavelength range, relative to a calibrated reference material of a pure (non-coated) glass slide.

Fourier transform infrared spectroscopy (FTIR) was performed in attenuated total reflection (ATR) mode on a Nicolet iS10 spectrometer (Thermo Fischer Scientific, Breda,

The Netherlands) equipped with a diamond crystal and via sampling at 4 cm^{-1} resolution, with 16 scans between wavenumbers of 4000 and 500 cm^{-1}.

3. Results

3.1. Mechanical Testing of Coatings

The mechanical properties of sprayed nanocellulose coatings strongly rely on the formation of a homogeneous and dense coating with internal hydrogen bonding, as evaluated previously [70]. However, the better mechanical properties of surface-modified nanocelluloses also indicated that additional crosslinking is needed for improved robustness and durability in practical applications [70]. Fulfilling the requirement to formulate nanocellulose coatings with enhanced mechanical resistance through crosslinking with BA, the macroscale mechanical properties of nanocellulose coatings with different configurations, layer thicknesses, and BA concentrations were evaluated. The multilayer (CNF/BA) coatings and mixed-layer (CNF + BA) coatings were deposited with 3, 5, and 10 spraying layers of nanocellulose, while evaluating the influences of coating configuration (Figure 1), low BA concentration in the interface layer or bulk of mixed-layer coatings (Figure 2), and high BA or ultra-high BA concentration in the bulk of multilayer or mixed-layer coatings (Figure 3).

The results for mechanical testing of multilayer versus mixed-layer coatings (Figure 1) were evaluated with optical micrographs after a tape test, rub test, cross-cut test without and with peeling (scoring grade 0 to 5), and scratch test under different normal loads. The reference coatings with pure nanocellulose (no BA) sprayed in 3, 5, and 10 layers were not further evaluated as they yield a low scratch resistance of <0.5 N. As a preliminary reference test for evaluating the role of pretreatment on the glass substrate, the coatings deposited onto bare glass substrates showed immediate adhesive failure: the presence of an adhesive interlayer of PDA and a first BA crosslinking layer on the glass/PDA/BA substrates was preferred to improve the wetting of the glass substrate and introduce adhesive bonding. For both multilayer (CNF/BA) and mixed-layer (CNF + BA) coatings, the high BA concentrations were initially used as a first crosslinking layer on top of the PDA adhesive interlayer on the glass interface, while simultaneously using the same high BA concentrations as a crosslinker in the bulk of the nanocellulose coatings. For the multilayer coatings with a high-BA crosslinker (CNF/high BA) (Figure 1, left), mechanical resistance is relatively weak with immediate mechanical failure for the multilayer coatings with 3 and 5 sprayed layers and somewhat better mechanical resistance for the multilayer coatings with at least 10 sprayed layers. However, the proposed multilayer coatings with a high-BA crosslinker did not provide a good combination of adhesive and cohesive strength, as failure immediately occurred after rub testing and scratch testing at 2 N for CNF/high BA (10 layers). For the mixed-layer (CNF + high BA) coatings with a high-BA crosslinker (Figure 1, right), better mechanical performance is noticed with a progressive increase in mechanical resistance with a higher number of sprayed layers, up to a maximum scratch resistance of 9 N (10 layers). The fracture pattern of the high-BA-crosslinked nanocellulose coatings is characterized by chevron lines behind the scratching track, which typically indicate a rather brittle failure of the coatings. These can intuitively be attributed to the formation of a densely crosslinked nanofiber network with low mobility of the single nanocellulose fibers (see below). Alternatively, the failure pattern of the coatings with weak mechanical resistance shows more deformation in the scratching track with the local upsetting of coating fragments in the scratching track and finally full coating removal. The local bulging of the coating implies interfacial failure in combination with local sliding effects of the coating over the glass substrate under scratching. The broad scratching tracks in the mechanically weak coatings are consequences of tearing apart of the loosely interlocked nanofibrils in the coating. In contrast with the presented coating configurations including intermediate drying of the subsequently deposited layers in multilayer or mixed-layer coatings, the coatings without intermediate drying between the different layers developed lower mechanical resistance (see Supplementary Figure S2). Therefore, intermediate drying after each sprayed layer better contributes to crosslinking for both multilayer and mixed-layer coatings. This could be explained by a higher degree of borate ester formation. The boric acid molecule needs to undergo two subsequent condensation reactions to form a stable crosslink between cellulose fibers, and water is produced as a byproduct [40]. When heat is applied it accelerates the loss of water molecules from the coating, and therefore the reaction equilibrium is pushed to completion. An incomplete reaction, perhaps in the absence of a drying step, would result in the instauration of weaker hydrogen bonds between the hydroxyl groups of the borate and cellulose. As an alternative method for deposition of the PDA/BA adhesive interlayer by dipping, the PDA and high-BA interface layers were sprayed and the coating performances were compared relative to the coatings with a single PDA interface layer without high BA (see Supplementary Figure S3); however, the lower coating adhesion in the absence of a high-BA interface layer (5 N scratch resistance, compared to 9 N) and lower mechanical properties of the coatings with a sprayed PDA/high-BA interface layer (8 N scratch resistance) were confirmed, as compared to a dip-coated high-BA interface later (9 N scratch resistance). The mechanical performance of the coating system indeed relies on the homogeneity (e.g., density) and crosslinking of the interface layer on glass as demonstrated by further chemical and physical characterization (see Section 3.4). While the present study focusses on the increase in robustness of coating systems with multilayers or mixed layers, the role of the single adhesive PDA or PDA/BA interface layer will be fully detailed in a separate study indicating negative influences due to higher roughness and porosity of a sprayed high-BA interface layer.

Starting from the same high BA concentrations as a crosslinker in the interface (glass/PDA/BA) and within the bulk of the CNF + BA mixed layers, the effects of low BA concentrations in either the interface or the bulk of the mixed-layer coatings were further evaluated (Figure 2). As a reference system, the preliminary adhesion trials with a single nanocellulose layer sprayed on pure PDA (no BA crosslinker in the interface) indicated immediate adhesive failure of the nanocellulose coating. Therefore, the synergistic effects between PDA and BA in the interface enhance the adhesion of the nanocellulose layers and can be explained by the formation of heteroleptic boron complexes with both the catechol units of PDA and the sugar residues from nanocellulose. Herein, the asymmetric borate ester acts as a linker between nanocellulose and the PDA layer, and is strongly adhered onto the glass surface. However, when a first layer of low BA concentration is applied in the interface in combination with a high BA concentration as a crosslinker in the bulk (Figure 2, left), tape test and cross-cut analysis indicate that the coating is more sensitive to adhesive failure in the interface with clear delamination as compared to the coatings with a high-BA interface layer. This corresponds with a reduced scratch resistance of 7 N and full removal of the coating from the substrate within the scratch track. Therefore, glass/PDA/BA interfaces with low BA concentration are not sufficient for good adhesion as compared to those with high BA concentration. While reducing the BA concentration as a crosslinker in the bulk of mixed-layer coatings, the coatings with mixed CNF + low-BA layers (Figure 2, right) present lower mechanical resistance than the coatings with mixed CNF + high-BA layers: the mixed CNF + low-BA coatings have a maximum scratch resistance of 6 N as compared to the scratch resistance of 9 N for mixed CNF + high-BA coatings. The mechanical failure of the coatings with low BA crosslinker concentrations is mainly attributed to cohesive failure, indicating a lower degree of crosslinking. As demonstrated from the tape test and cross-cut test, however, the adhesive properties of the mixed-layer

(CNF + low BA) coating with 10 layers are maintained with a high-BA interface layer. Although the internal strength of the coatings with low-BA crosslinker concentrations in the bulk fail, the good adhesive properties remain controlled by the high BA concentration in the interface. As such, mechanical resistance in both the bulk and the interface of the coating are separately controlled by the BA concentration.

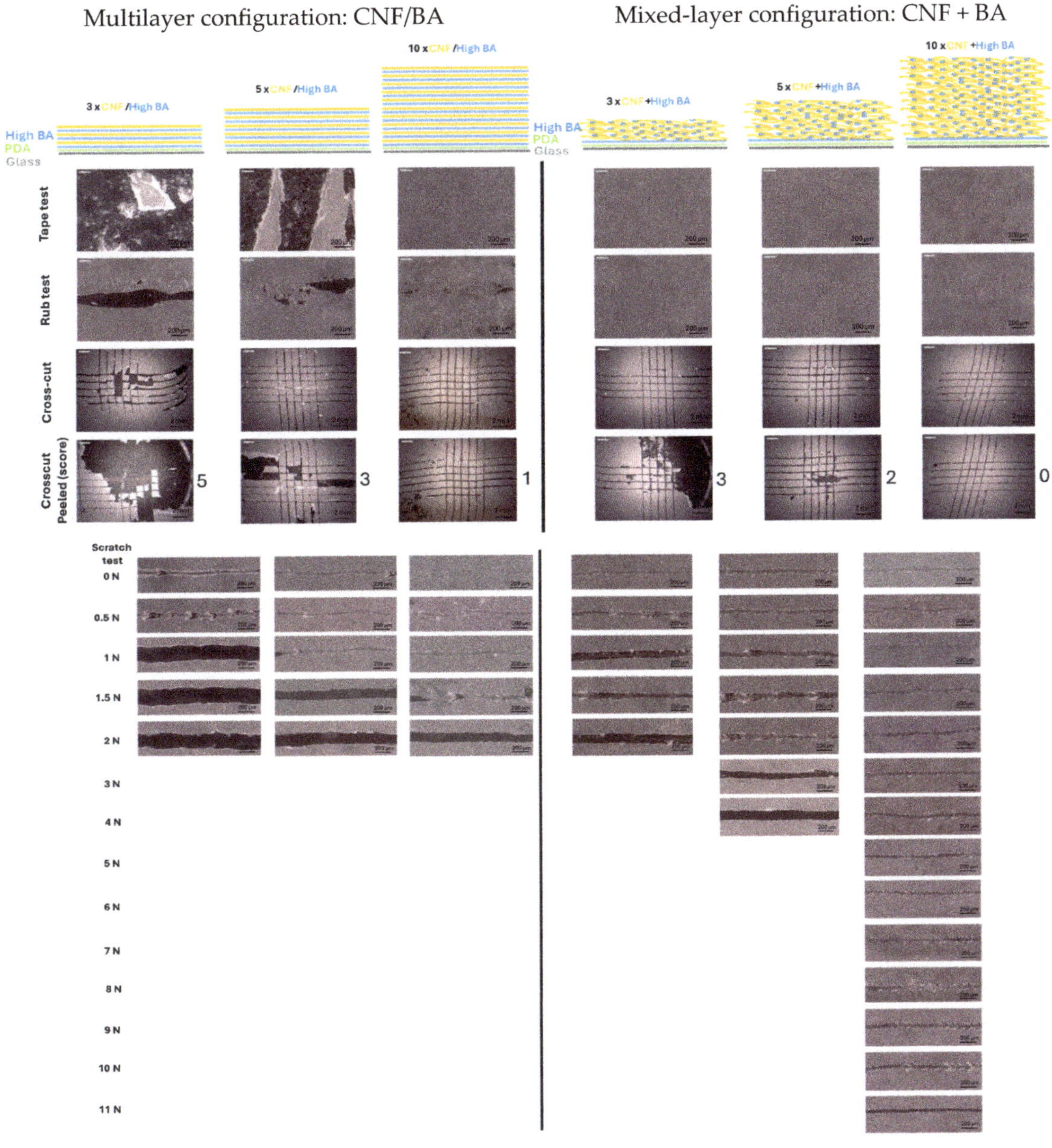

Figure 1. Mechanical resistance testing of multilayer (CNF/BA) coatings (**left**) and mixed-layer (CNF + BA) coatings (**right**), with the same high-BA crosslinker concentrations (10 mM) in the interface layer and coating bulk.

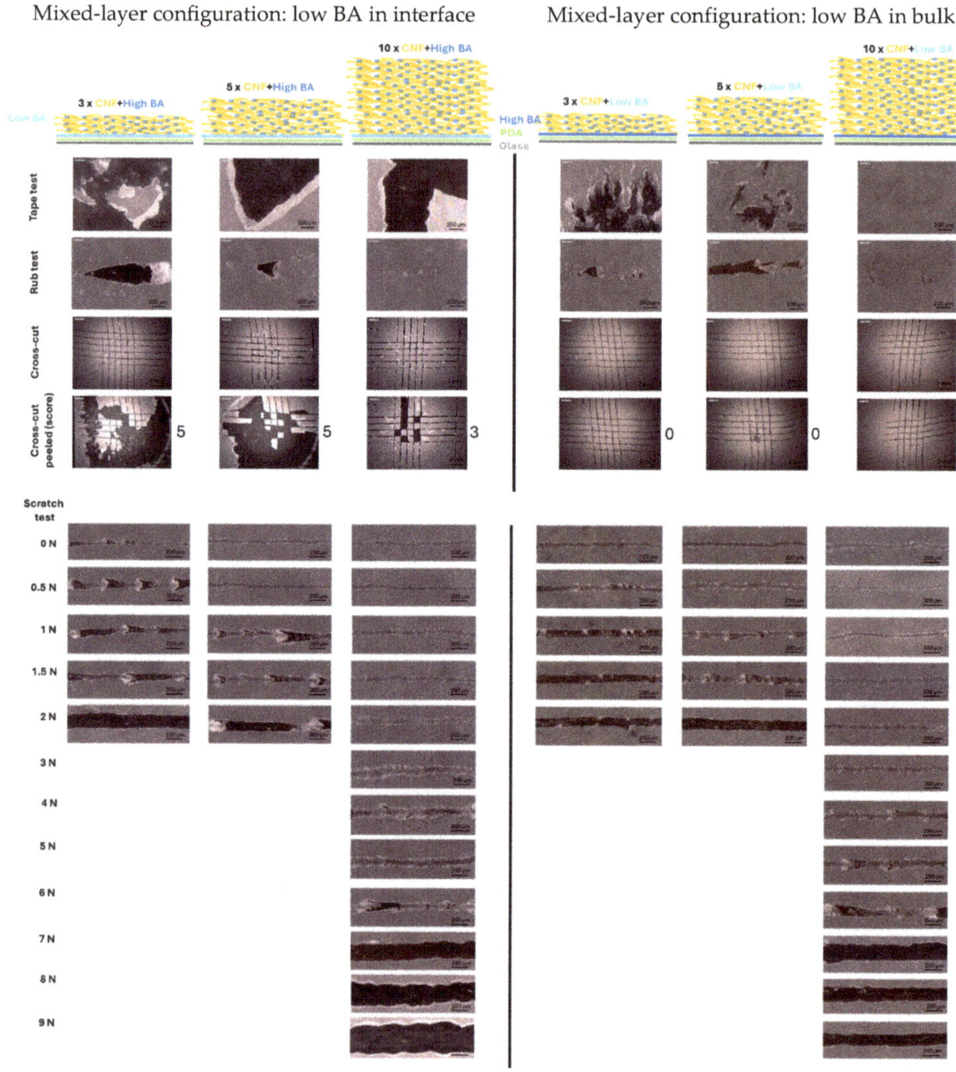

Figure 2. Mechanical resistance testing of mixed-layer (CNF + BA) coatings with low-BA crosslinker concentrations (1 mM) either in the interface layer (**left**) or in the coating bulk (**right**).

Finally, the mechanical properties with ultra-high-BA crosslinker concentrations in the bulk of the coatings were evaluated for multilayer (CNF/ultra-high BA) and mixed-layer (CNF + ultra-high BA) coatings (Figure 3). Also, the application of an ultra-high BA concentration in the adhesive interlayer glass/PDA/BA was investigated, but it immediately caused degradation of the PDA layer. This effect can be associated with a strong interaction of the catechol units with boric acid, which was reported as an effective inhibitor of dopamine polymerization and PDA adhesion [71]. Dip coating with the ultra-high BA concentration caused substantial leaching from the PDA layer, observed as discoloration of the brown coating and complete detachment after a few hours. For this reason, the gradient coating configurations were designed with the initial layers with high BA concentrations in contact with the glass/PDA interface layer, while avoiding direct contact between the glass/PDA interface layer and the ultra-high BA concentrations. In combination with a high-BA interface layer, a number of intermediate high-BA layers were applied in both multilayer coatings and mixed-layer coatings. For the multilayer coatings (Figure 3, left), a single high-BA interface layer in contact with the glass/PDA adhesive layer efficiently provides good mechanical adhesion with the substrate. The cohesive strength of the coating under scratch testing has improved towards 8 N for multilayer coatings with CNF/ultra-high BA in contrast with the scratch resistance of 1.5 N for multilayer coatings with CNF/high BA. Adversely, for the mixed-layer coatings (Figure 3, right), good adhesive strength is provided through the high-BA interlayer while the cohesive strength is reduced towards a maximum scratch resistance of 6 N for the mixed-layer (CNF + ultra-high BA) coatings in contrast with the scratch resistance of 9 N for the mixed-layer (CNF + high BA) coatings. This can intuitively be attributed to the very dense crosslinking of the mixed-layer (CNF + ultra-high BA) coatings with reduced mobility of the entangled fiber network and high brittleness, as visually observed by the chevron lines in the scratching track. In contrast, the mixed-layer (CNF + high BA) coatings with the highest scratch resistance show a smooth scratching track related to a ductile failure mode.

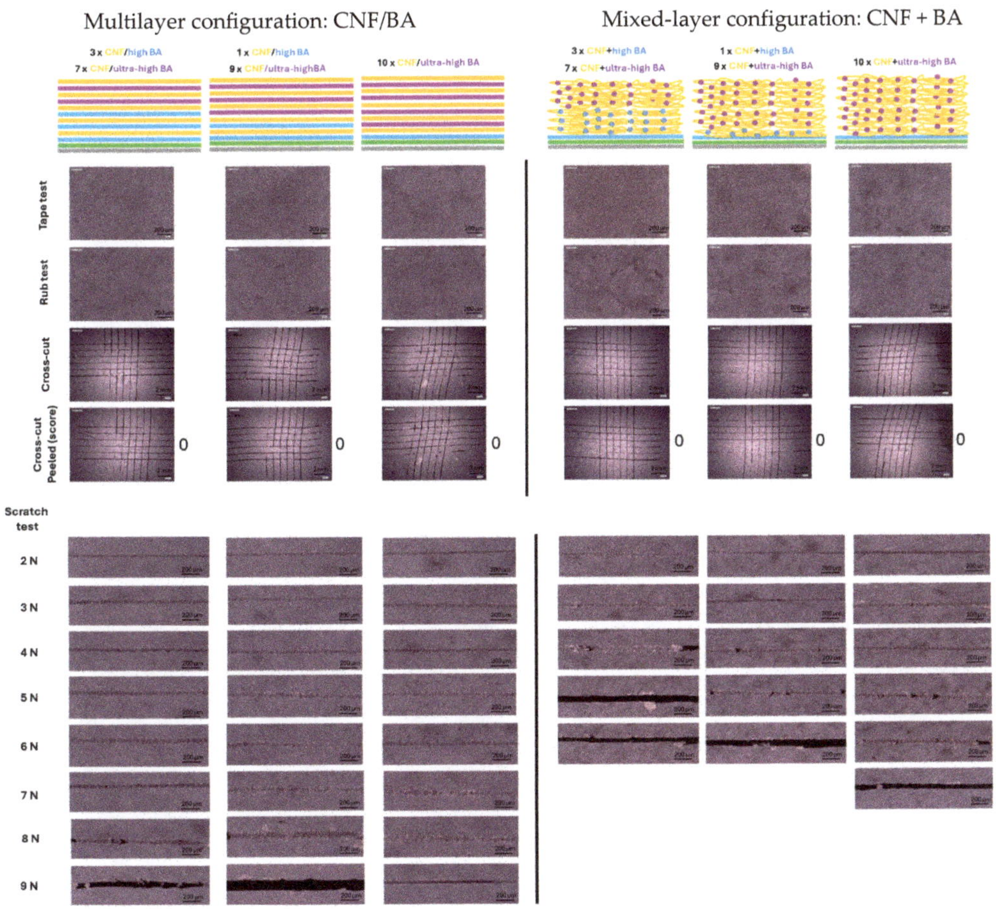

Figure 3. Mechanical resistance testing of multilayer (CNF/BA) coatings (**left**) and mixed-layer (CNF + BA) coatings (**right**) with ultra-high-BA crosslinker concentrations (100 mM) in the interface layer and coating bulk, applied in combination with the high-BA crosslinker concentration (10 mM), creating gradient coating layers.

In conclusion, a balance between cohesive and adhesive failure for dried coatings with enhanced mechanical robustness has been experimentally determined through controlling the BA crosslinker concentration in the glass/PDA/BA interface and bulk of the coating. Depending on the coating configuration, the multilayer (CNF/ultra-high BA) coatings and mixed-layer (CNF + high BA) coatings show the best cohesive properties, while an interface layer with high BA concentration is required for enhanced adhesion. The chemical interactions in relation to qualitative data for crosslinking, confirming the enhanced mechanical properties, are further discussed in the following paragraphs.

3.2. Morphology and Surface Roughness

The morphology and topography of sprayed nanocellulose coatings with high BA crosslinker concentrations in different configurations are illustrated in Figure 4, including the multilayer (CNF/high BA) coatings (Figure 4a), and mixed-layer (CNF + high BA) coatings (Figure 4b) deposited with intermediate drying or without intermediate drying (wet). Within a homogeneous coating region of fiber deposits, the microscopic images at different magnifications (×600, ×1200) show a regular arrangement of CNF nanofibrils and formation of a densely packed nanofiber coating with an intrinsic nanoscale pattern related to the organization of the nanofibrils. The 2D and 3D topographical images at different magnifications (×20, ×50, ×600) show the progressive formation of fiber entanglements and interactions as the coating thickness increases. With an increasing number of sprayed layers, there is a tendency for fiber agglomeration and formation of protrusions at the surface to occur.

The values for average surface roughness Sa were determined from the topographical surface scans at a magnification of ×20 or ×50, indicating the differences between multilayer (CNF/high BA) coatings and mixed-layer (CNF + high BA) coatings (Figure 5). The changes in surface roughness for different coating configurations and coatings with different numbers of layers are consistent for both magnifications, demonstrating a good representation of the coating morphologies at different scale levels. The average surface roughness Sa obviously increases progressively with a higher number of sprayed layers, while it is consistently higher for the coatings with intermediate drying (as compared to a wet deposited layer), and lower for the mixed-layer coating layers (as compared to the multilayers). The results indicate that the multilayer coatings are more heterogeneous as compared to the mixed-layer coatings, while additional structural organization or crosslinking during the intermediate drying steps may affect the arrangement of the nanofibrils, resulting in a higher surface roughness. The lower surface roughness of the mixed-layer coatings compared to multilayer coatings may suggest fewer fibril protrusions at the surface and a lower porosity, in parallel with a higher crosslinking density. Also, the intermediate evaporation of water during intermediate drying may introduce local re-arrangements and organization of the nanofibrils. As compared to the sprayed coatings of CNF (no BA crosslinker) in three or five layers [70], the values for average surface roughness are in the same range, between Sa = 0.2 μm (three-layer CNF) and Sa = 0.4 μm (five-layer CNF). The roughness of sprayed CNF coatings shows good agreement with the roughness of mixed-layer (CNF + high BA) coatings, while the higher roughness of multilayer (CNF/high BA) coatings indicates their more heterogeneous morphology.

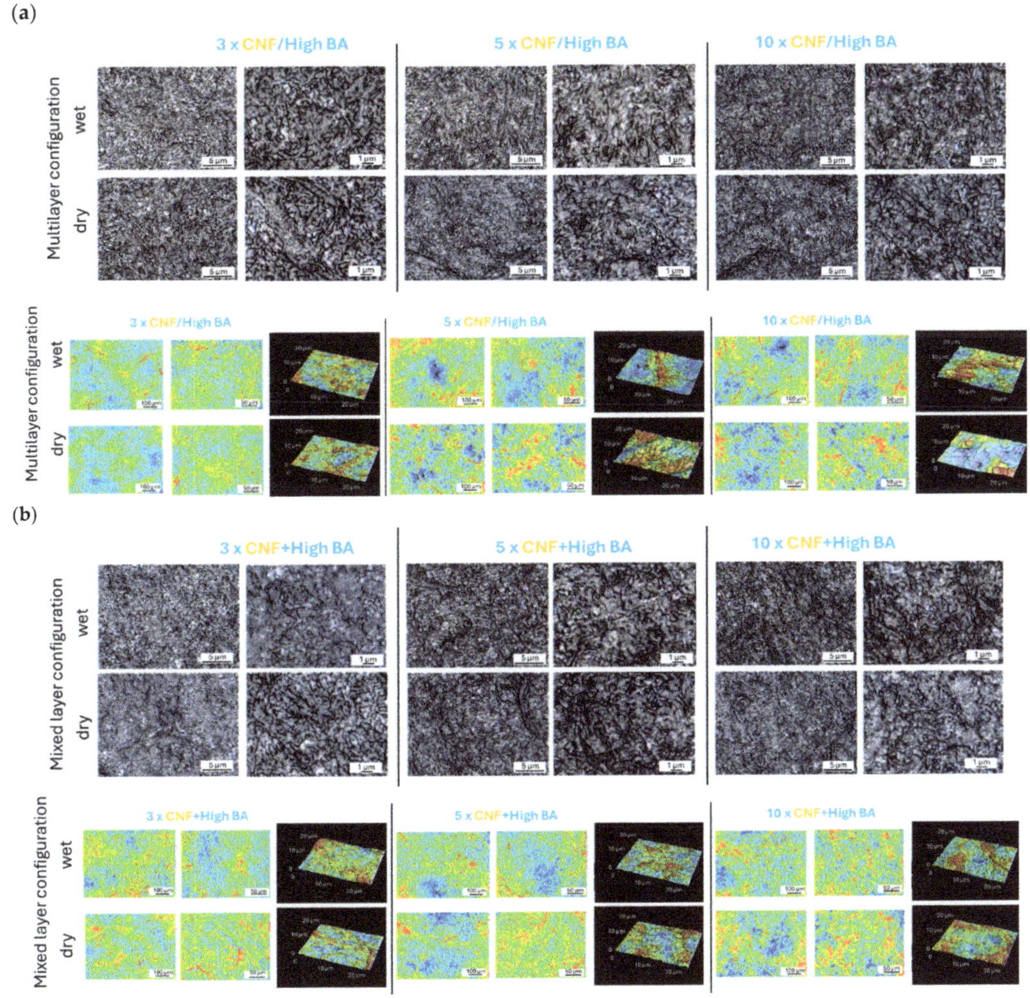

Figure 4. Morphology and topography images obtained from confocal laser scanning microscopy for different coating configurations deposited under wet conditions (no intermediate drying) or dry conditions (with intermediate drying) with different numbers of subsequently sprayed layers in (**a**) multilayer (CNF/BA) coatings and (**b**) mixed-layer (CNF + BA) coatings. Color figures represent z-scale from blue (0 μm) to red (5 μm).

The more detailed observations of local coating morphologies at different magnifications, as shown in Figure 6, indicate the presence of organized crystalline structures in multilayer coatings (both CNF/high BA and CNF/ultra-high BA). Depending on the drying conditions, the crystallite shapes are more spherical for the wet layer deposits, while they have further developed into dendrite shapes after intermediate drying. The more specific details on different crystallite structures for multilayer coatings with or without intermediate drying are shown in Supplementary Figure S4. The crystallite formation under drying of BA from the aqueous phase has been described before and results in similar platelet structures, where the crystals are formed by dehydration under evaporation and the crystallite size depends on the concentration, dispersant, temperature, and evaporation rates of BA [72]. If present, a secondary phase such as nanocellulose may either act as a nucleator or a crystal growth inhibitor during the crystallization of BA, depending on its miscibility and surfactant properties [73]: in particular, BA crystals with a hexagonal shape were obtained in combination with other solvents and have the most developed shapes after drying at around 70 °C in the presence of carbohydrates [74]. For mixed-layer (CNF + BA) coatings, no crystallization effects were observed as the nanofiber network likely hinders crystal growth, resulting in more homogeneous coating morphologies. In relation to the observed coating performance, the crystallite formation in multilayer (CNF/BA) coatings obviously weakens the mechanical properties as it might introduce brittleness or reduce the interactions between BA and CNF. The formation of BA crystals as contaminants also occurred previously under conditions of relative humidity above 15% and resulted in poor adhesion properties of multilayer packaging due to the increase in softening temperature [75].

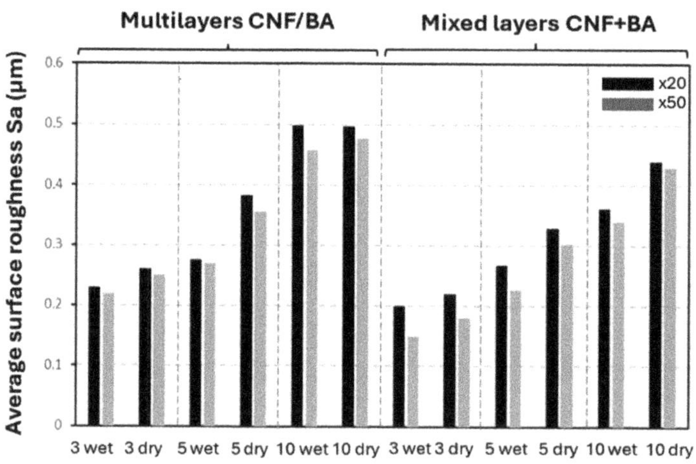

Figure 5. Average surface roughness values Sa for multilayer (CNF/BA) and mixed-layer (CNF + BA) coatings, with different numbers of subsequently sprayed layers deposited under wet conditions (no intermediate drying) or dry conditions (with intermediate drying).

The coating thickness and density for multilayer and mixed-layer coatings are illustrated in Figure 7, as determined from experimental weight and thickness measurements. There is a linear increase in coating weight with a higher number of sprayed layers (Figure 7a) for both the multilayer (CNF/high BA) and the mixed-layer (CNF + high BA) coatings. The repetition of the measurements is confirmed through good overlap in weight data for the coatings with 3, 5, and 10 sprayed layers and intermediate drying for both multilayer and mixed-layer coatings. The weight of multilayer (CNF/high BA) coatings is somewhat higher as compared to the mixed-layer (CNF + high BA) coatings, likely because of the lower control on subsequent spraying of the two separate layers of CNF and high BA with a slight overload of BA (higher than the 1:1 ratio as compared to the mixed-layer coatings). Otherwise, the increase in weight of the wet deposited multilayer and mixed-layer coatings does not follow a consistent trend, deviating from a linear increase (mainly at a higher number of sprayed layers) due to an overspray with partial removal of the wet coating layer upon subsequent spraying. As opposed to the weight measurements, however, the coating thickness for the mixed-layer (CNF + high BA) coatings is higher than that for the multilayer (CNF/high BA) coatings (Figure 7b). Therefore, the calculated coating densities (Figure 7c) indicate higher densities for the multilayer (CNF/high BA) coatings, and lower densities for the mixed-layer (CNF + high BA) coatings. As a function of BA crosslinker concentrations in the bulk or interface layer (Figure 7d), the coating densities were calculated from coating weight and thickness measurements. In comparison, the coatings with a sprayed PDA/BA interface layer on glass demonstrate inferior properties and were not further considered (see Supplementary Figure S5); in particular, the low density of the coatings with a sprayed interface layer indicates a higher heterogeneity, in line with the lower mechanical resistance of these coatings as observed before.

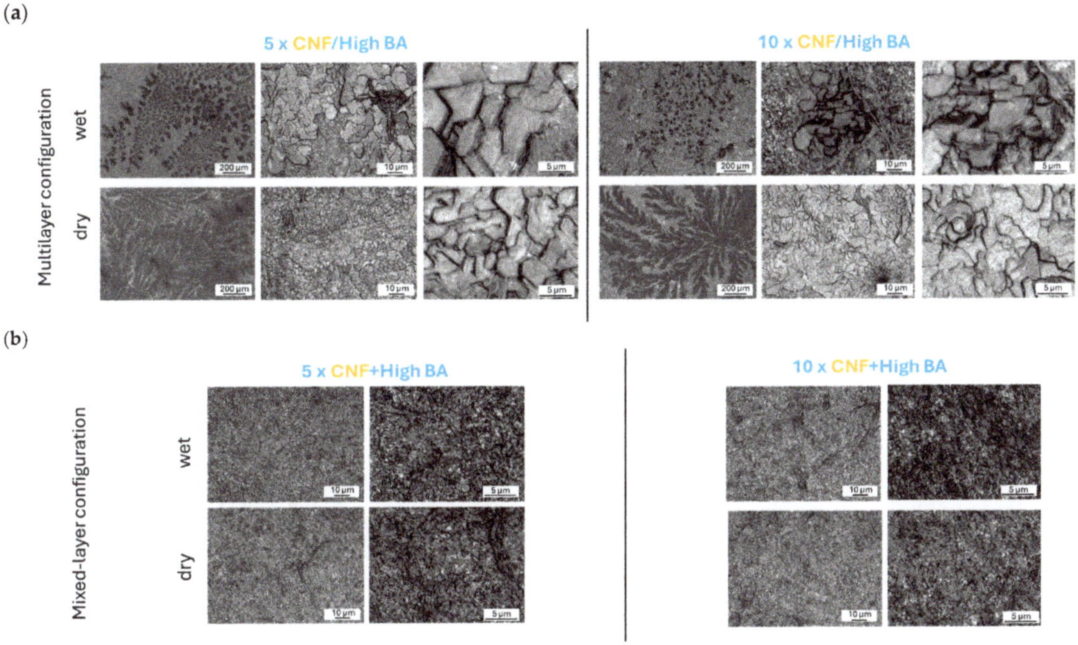

Figure 6. Detailed morphological evaluation of coating homogeneity/heterogeneity, indicating (**a**) the local formation of BA crystallites in the case of multilayer (CNF/BA) coatings, and (**b**) homogeneous fibrous morphology in the case of mixed-layer (CNF + BA) coatings.

In parallel with previous data for spraying of CNF coatings [69], the coating density increases with the number of sprayed layers and stabilizes at above five to six sprayed layers due to the presence of local heterogeneities for the thinner coatings. For the multilayer coatings, densities are higher as compared to the mixed-layer coatings, and the increase in crosslinker concentration introduces an increase in coating density, shown by comparing the multilayer coatings (CNF/high BA and CNF/ultra-high BA). The higher density for multilayer coatings can be attributed to formation of crystallites, as demonstrated in the previous paragraph. For the mixed-layer coatings, the higher BA crosslinker concentrations also involve an increase in coating density by comparing mixed-layer CNF + low-BA and CNF + high-BA coatings, which confirms the crosslinking in the bulk of the coatings.

In relation to the reported mechanical performance of multilayer and mixed-layer coatings, the coating configurations with the highest density do not uniquely introduce the best mechanical resistance when comparing multilayer (CNF/BA) and mixed-layer (CNF + BA) coatings. Rather, the high density of multilayer (CNF/BA) coatings might be related to the formation of BA crystals that negatively affect the mechanical properties of the coating through brittleness and reduced chemical interactions between the BA crosslinker and nanocellulose (see Section 3.4). Alternatively, when comparing the mixed-layer (CNF + BA) coatings at different BA concentrations, the coatings with high-BA crosslinker concentrations provide the highest density and best mechanical resistance (see Section 3.4).

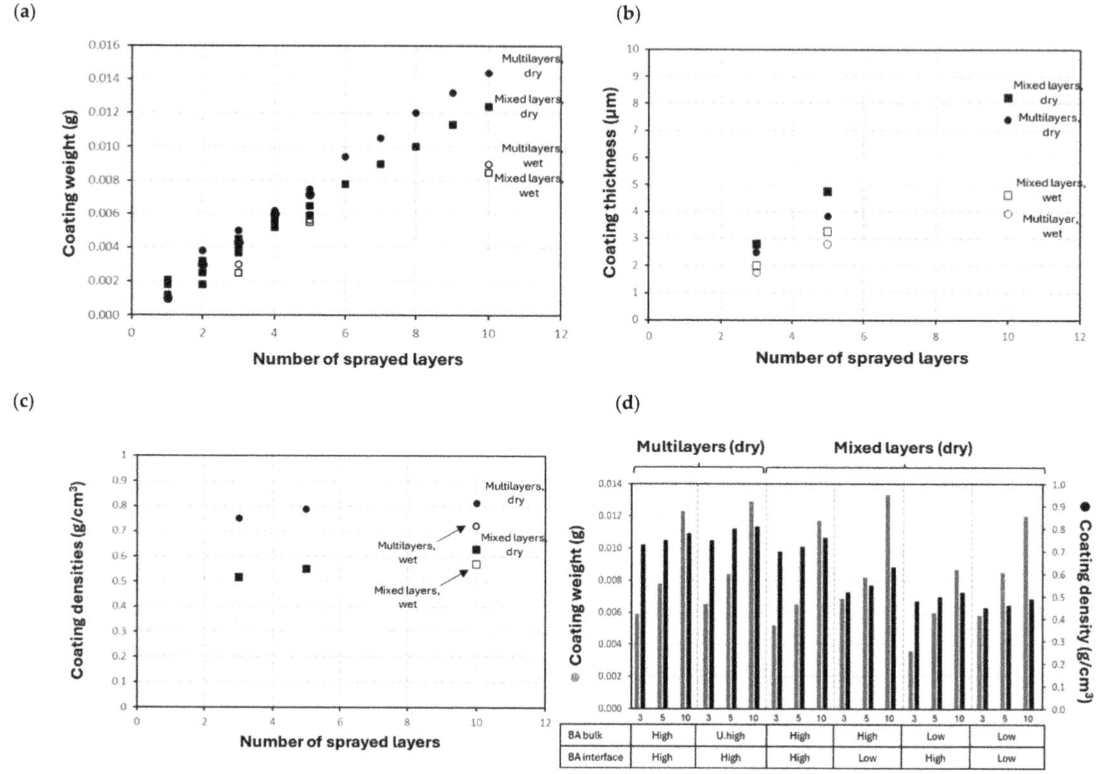

Figure 7. Coating deposition characteristics as a function of the number of sprayed layers and final characteristics, including (**a**) coating weight, (**b**) coating thickness, and (**c**) coating densities. (**d**) Coating weight and coating density summary for different coating configurations with 3, 5, and 10 sprayed layers.

3.3. Physical Coating Properties

While the nanocellulose materials are strongly hydrophilic, their structural and dimensional stability in aqueous environments may be improved by crosslinking [13]. Hence, the effects of crosslinking of the nanocellulose coatings on the stability and interaction of water droplets deposited onto the surface were investigated by static water contact angle measurements to observe the spreading of the water droplets depending on the coating configuration and crosslinking (Figure 8).

The decrease in water contact angles over an observation time of 20 s was noticed because of spreading and penetration of the water on the coating (Figure 8a). The reference measurements of water droplets spreading on the substrates (glass, glass/PDA, and glass/PDA + BA) are presented in Supplementary Figure S6. The measurements for pure CNF coatings could not be performed as the water directly penetrates into the coating, while the crosslinked CNF coatings show retarded penetration of the water. The initial water contact angles upon direct contact (time < 1 s) for crosslinked coatings may increase due to roughness effects, but they are particularly higher for the mixed-layer coatings (CNF + BA), showing better water resistance. A direct relation between surface roughness and the initial water contact angle is not evident for the strongly hydrophilic coatings. It is obvious that the water stability over time increases for the coatings with a higher number of sprayed layers. For multilayer (CNF/BA) coatings, the rapid spreading of a water droplet over the coating is observed with a fast decrease in the water contact angle value over time. The high roughness of the multilayer (CNF/BA) coatings possibly relates to a higher surface porosity and easy water penetration. The multilayer coatings with intermediate drying and a high number of sprayed layers become more stable than the coatings deposited without intermediate drying, as the mechanical resistance of the coatings with intermediate drying also improved. For the mixed-layer (CNF + BA) coatings, the spreading and penetration of water over time were retarded with a higher stability for the mixed-layer coatings with high BA concentrations in the bulk and/or interface. The improvement in water stability is mostly determined by the high BA concentration as a crosslinker in the bulk (as related to the penetration of water in the bulk of the coating), while there additionally is a minor influence of the BA concentration in the interface (as related to the penetration of water in the interface). The stability and higher water contact angles for BA-crosslinked nanocellulose coatings are better as compared to previous studies reporting lower water contact angles for BA-crosslinked nanocellulose coatings on wood [56]. In parallel with the mechanical resistance of the mixed-layer (CNF + high BA) coatings, the coatings with high BA concentrations in the bulk or interface have higher water stability than the coatings with low BA concentrations, as related to the crosslinking density of the coatings. The fast wetting on multilayer (CNF/high BA) coatings can be seen from the progress of the water contact line (Figure 8b), while slow wetting on mixed-layer (CNF + high BA) coatings is observed as a slow advance of the water contact line over the coating and sharp definition of the water droplet border (Figure 8c). During drying of the water droplet, the contact line progressively retracts and the nanofibers re-arrange into a closed and dried coating (Figure 8d); the latter is an indication of structural recovery of the mixed (CNF + BA) coatings after water contact.

Although the thin nanocellulose coatings provide optical transparency owing to the size features of the cellulose nanofibers below the wavelength of light, the non-uniformities in the coatings, partially due to effects such as local organization of fibers into microscale entities, surface roughness, or porosity, can result in light-scattering effects and haze [76]. Therefore, the transparency loss of the coatings depending on the number of sprayed layers and internal crosslinking was quantified for the different coating configurations relative to the uncoated glass slide, both providing performance data and qualitatively relating to the internal homogeneity of the coating structure. The raw data of UV-VIS spectra and visual observation with an optical microscope are illustrated for some coating configurations (see Supplementary Figure S7), and (relative) transparency values at a selected wavelength (600 nm) are summarized for mixed and multilayer coatings with different BA concentrations as crosslinkers in the bulk or interface layer (Figure 9a). For both multilayer (CNF/BA) and mixed-layer (CNF + BA) coatings, the high concentrations of BA crosslinkers in the bulk of the coating provide higher transparency as compared to the low BA concentrations. Otherwise, the mixed-layer (CNF + high BA) coatings provide the highest transparency, reflecting better homogeneity as compared to the multilayer (CNF/BA) coatings. The transparency cannot solely be related to the coating thickness, but it relates best to the coating densities, with significant influence of BA crosslinker concentration in the bulk (Figure 9b). The higher density of the coatings through crosslinking intuitively implies the close interaction between the fibers, and hence there are likely fewer pores for light scattering. In comparison, the transparency of coatings with a sprayed PDA/BA interface layer on glass was inferior (see Supplementary Figure S8), in parallel with their heterogeneity and lower mechanical performance as indicated before, and this configuration was not further considered. As compared to the literature [77], a high transparency of >90%

was only obtained for nanometer thickness of nanocellulose composite coatings with high density and a low fraction of nanovoids.

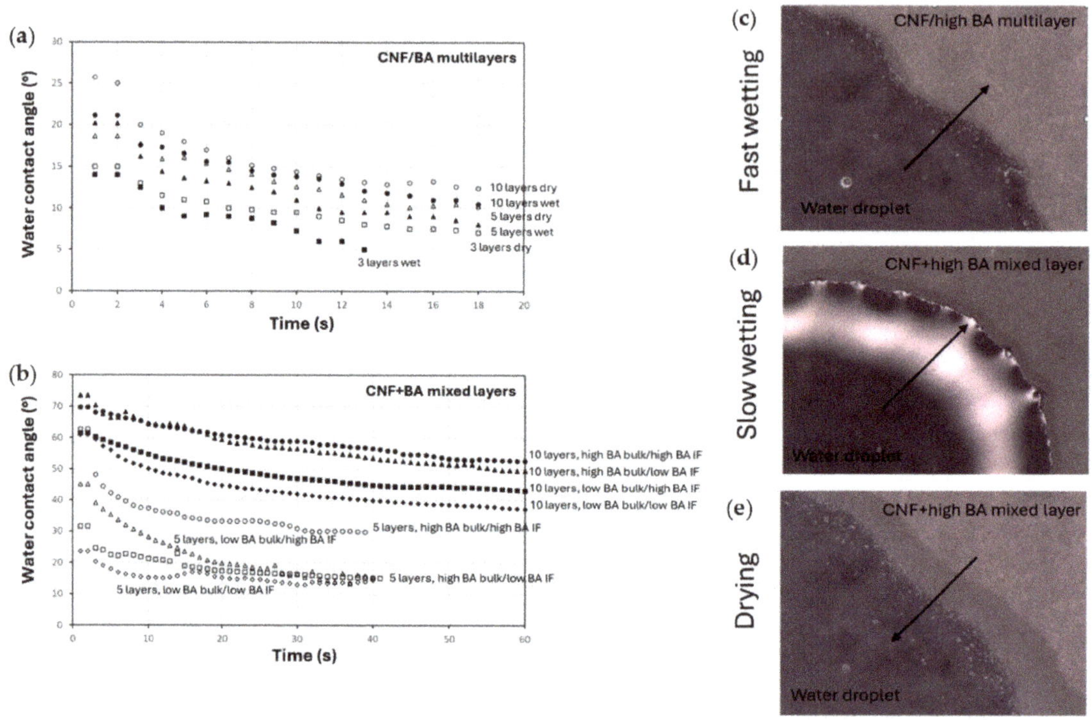

Figure 8. Water interaction of crosslinked coatings as determined from water contact angle measurements over time for (**a**) multilayer (CNF/BA) coatings and (**b**) mixed-layer (CNF + BA) coatings. Illustrations of the water droplet contact line with (**c**) advancing of a water droplet onto a multilayer (CNF/high BA) coating, (**d**) advancing of a water droplet onto a mixed-layer (CNF + high BA) coating, and (**e**) receding of a water droplet on a mixed-layer (CNF + high BA) coating with recovery of the coating structure during drying.

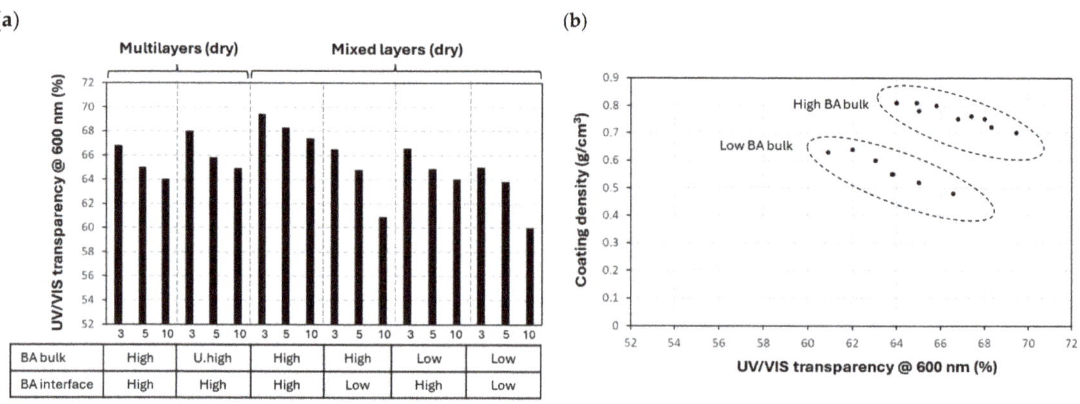

Figure 9. Transparency of crosslinked coatings as determined from UV/VIS spectroscopy: (**a**) transparency data summary for different coating configurations with 3, 5, and 10 sprayed layers, (**b**) relation between transparency and coating densities.

3.4. Chemical Coating Properties

The chemical interactions between the CNF and BA in different coating configurations were studied by FTIR spectroscopy, revealing the effects of intermediate drying, the number of sprayed layers, and BA concentrations (Figures 10–12). The characteristic spectral bands for boric acid are located at around 1400 cm^{-1} (B-O, B-O-B asymmetric stretching) and 1150 cm^{-1} (B-OH in-plane bending), which both strongly overlap with the characteristic fingerprint region of cellulose presenting the C-H and O-H bending bands (1450, 1426, 1365, 1335, 1314, 1280, 1247, 1203, and 1159 cm^{-1}) and C-O vibration bands (1104, 1052, 1028, 1002, and 985 cm^{-1}). The same bands in the range of 1420 to 1407 cm^{-1} were observed due to asymmetric B-O-C vibrations in the nanocomposite films with cellulose nanocrystals and BA, attributed to hybrid film formation [78]. Changes in the cellulose region at 1055 cm^{-1} and 1029 cm^{-1} were observed for some films of microcrystalline cellulose with a BA crosslinker, but no clear indication of interactions within this region was given as it might rather relate to the changes in cellulose structure [52]. Therefore, the maximum intensity of cellulose bands was used as an internal reference for our study, and the BA spectral bands can be related to the content of boric acid.

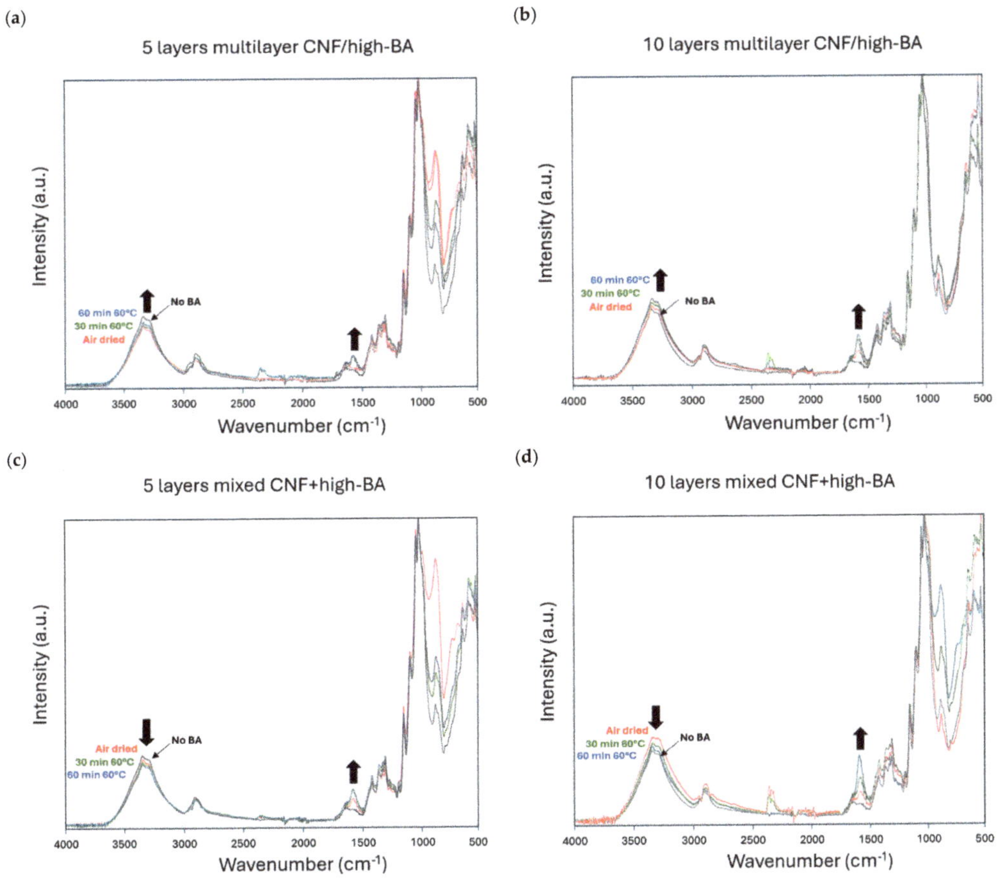

Figure 10. Infrared spectroscopy indicating the effects of drying on free hydroxyl groups for different coating configurations, including (**a**,**b**) multilayer (CNF/BA) coatings and (**c**,**d**) mixed-layer (CNF + BA) coatings. Arrows point towards interesting intensity changes (increasing or decreasing).

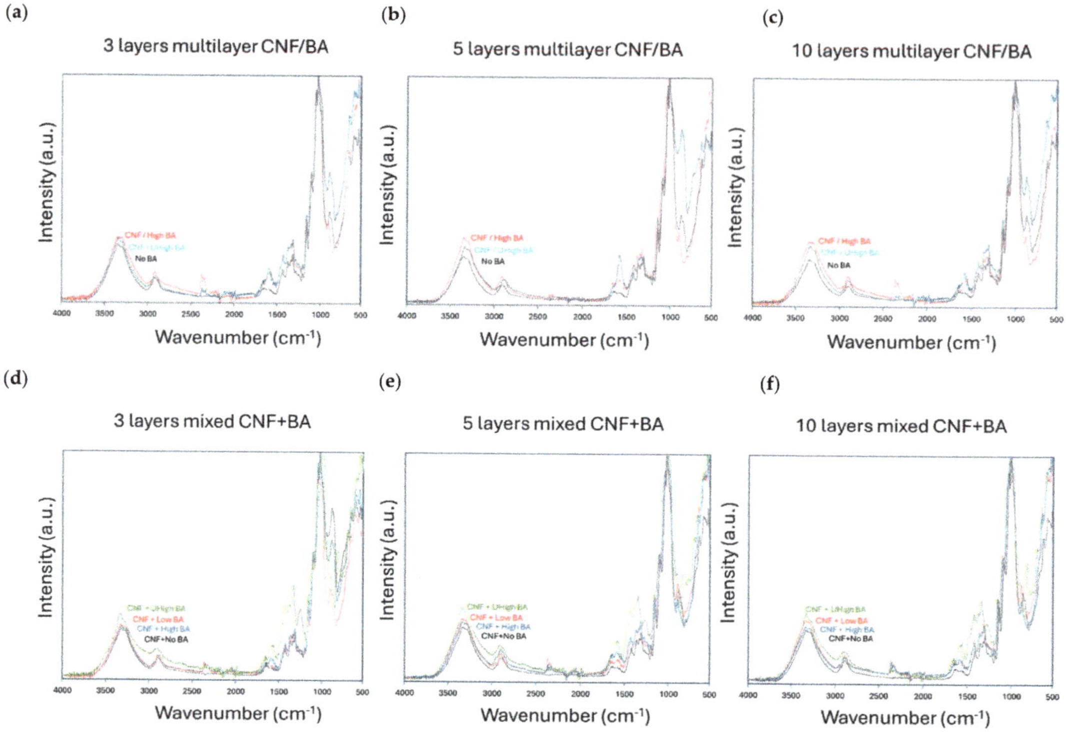

Figure 11. Infrared spectroscopy indicating the effects of BA concentration in the bulk for different coating configurations, including (**a–c**) multilayer (CNF/BA) coatings and (**d–f**) mixed-layer (CNF + BA) coatings.

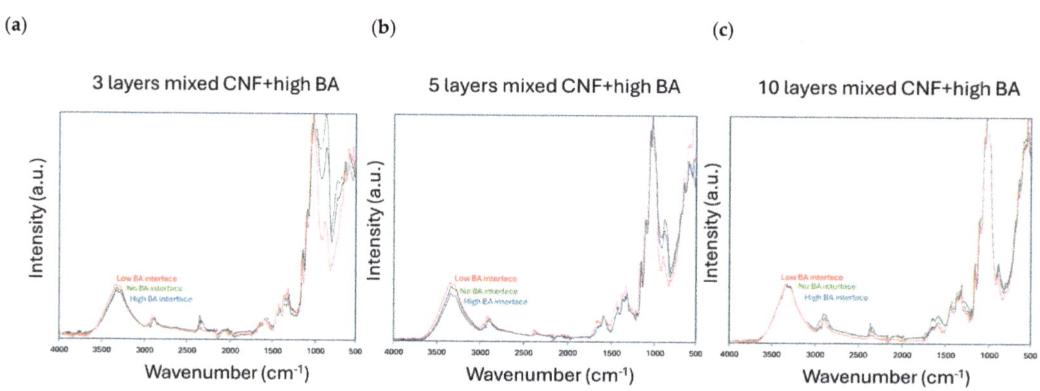

Figure 12. Infrared spectroscopy indicating the effects of BA concentration in the glass/PDA/BA interface layer for mixed-layer (CNF + high BA) coatings with different numbers of layers, including (**a**) 3 mixed layers, (**b**) 5 mixed layers, and (**c**) 10 mixed layers.

The spectral bands at 3800 to 3000 cm^{-1} contain complex contributions of O-H stretching for both cellulose and boric acid, with a maximum at around 3330 cm^{-1} for CNF and at around 3320 cm^{-1} for BA. While both peaks representing free hydroxyl cannot be separately resolved, they contribute to the overall intensity and width of the spectral hydroxyl band region and give a qualitative indication of relative differences in interaction between the components depending on the coating configuration. The same variations in intensities for this specific spectral band region were observed after the chemical modification of poplar wood flour with boric acid [79]. The interactions between free hydroxyl groups of BA and CNF are indicative of the crosslinking reaction forming a dense network through hydrogen interaction and possible covalent bonding under evaporation of the water. Depending on the composition and presence of hydroxyl groups in both components, an increase in free hydroxyl content is expected when mixing under conditions without interaction between the components. Alternatively, a reduction in the free hydroxyl content would imply the interaction and bonding between both components. According to the literature observations for CNF and CNC nanocellulose films [80], a proportional decrease in spectral bands of the nanocellulose can be expected with an increase in the BA ratio together with new peaks of BA.

For multilayer (CNF/high BA) and mixed-layer (CNF + high BA) coatings, the effects of drying are illustrated by comparing the spectra for coatings deposited under wet conditions (no intermediate drying) and dry conditions (with intermediate drying), as shown in Figure 10. The different effects of drying on the free hydroxyl band intensities (3800 to 3000 cm^{-1}), relative to an increase in band intensities related to pure BA (1400 cm^{-1}), are noticed for the conditions of air drying (no oven), mild drying (30 min, 60 °C), and intense drying (60 min, 60 °C). For multilayer coatings, the OH band intensity progressively increases depending on the more intense drying conditions for both thin (3- to 5-layer) and thick (10-layer) coatings, with the higher intensities related to the higher coating thickness. These trends signify no reduction in free hydroxyls due to unfavorable interactions or bonding in multilayer coatings: the results for weak chemical interaction are indeed in line with their relatively weak mechanical properties. The increase in band intensity for drying of the coatings with the same number of layers may be related to possible migration of free species towards the surface. When increasing the concentration of high BA towards ultra-high BA for the multilayer CNF/BA coatings, the reduction in OH band intensities upon drying indeed confirms the crosslinking (see Supplementary Figure S9), in parallel with the better mechanical performance. On the contrary, the drying of mixed-layer coatings indicates a progressive reduction in OH band intensities depending on the more severe drying conditions. The intense mixing of BA and CNF may indeed provide better conditions for crosslinking through the entire bulk of the mixed-layer coating at high BA concentrations. Alternatively, the interactions between BA and CNF in the multilayer coatings are restricted to the interfaces between both components, and the ultra-high BA concentrations are needed to ensure crosslinking. In conclusion, the first interpretation of interactions around free hydroxyl groups depending on drying conditions and coating configurations provides consistent data despite the relatively low concentrations of BA in the overall coating composition.

The chemical effects and interactions of variable BA crosslinker concentrations in the bulk of multilayer coatings (CNF/BA) and mixed-layer coatings (CNF + BA) are further detailed in the spectra for coatings with different numbers of layers, as shown in Figure 11. The spectra for multilayer coatings show consistent trends for all coatings with 3, 5, and 10 layers (Figure 11a–c), showing that the OH band intensities reduce for ultra-high-BA crosslinker concentrations as compared to high-BA crosslinker concentrations. This is a final confirmation that the multilayer coatings with ultra-high BA concentrations introduce better crosslinking as compared to the multilayer coatings with high BA concentrations, in parallel with the observed improvement in mechanical properties. The spectra for mixed-layer coatings with 3, 5, and 10 sprayed layers (Figure 11d–f) indicate that the OH band intensities are lowest for the compositions with high BA concentrations, which is proof of better crosslinking as compared to the composition with ultra-high BA concentrations. The higher intensities of free hydroxyl bands for the mixed-layer coatings with ultra-high BA concentrations can indeed be related to the reduced crosslinking interactions in parallel with the weaker mechanical properties observed before.

The chemical effects of the BA crosslinker concentrations in the interface of mixed-layer coatings (CNF + high BA) with the glass/PDA/BA substrate are illustrated in the spectra of Figure 12 for coatings with different thicknesses. The OH band intensities are compared for the coatings with 3, 5, and 10 mixed layers deposited on glass/PDA (no BA) and glass/PDA/BA substrates with low BA or high BA concentrations in the interface. The most differences in free hydroxyl group interactions are obviously observed for the thin coatings (3 or 5 layers), and no effects are seen for the thickest coatings (10 layers), as the interface interactions are not detected in the latter case. The relative changes in band intensities related to free hydroxyls are similar for the coatings with three or five sprayed layers and somewhat more pronounced in the latter case due to the higher thickness of coating present while still sensing the interactions with the interface layer. The relatively low spectral band intensities related to free hydroxyl groups for coatings deposited onto the glass/PDA/BA substrate with high BA concentrations as compared to low BA concentrations reflect a reduction through crosslinking between the mixed-layer coating and the interface with high BA concentrations, while the high concentrations of free hydroxyl groups for coatings deposited onto the interfaces with low BA concentrations apparently do not introduce the appropriate crosslinking. These observations were confirmed by previous mechanical testing, where the low BA concentration in the interface for mixed-layer coatings caused premature failure. In conclusion, the qualitative trends observed in interactions between BA and CNF through a reduction in free hydroxyl groups after crosslinking for given coating configurations are confirmed by the enhancement in mechanical properties of the coatings.

4. Conclusions

The mechanical robustness of sprayed nanocellulose coatings was enhanced through the design of different configurations with multilayer and mixed-layer coatings, including different concentrations of boric acid (BA) as an internal crosslinker and deposition on glass substrates with an adhesive interlayer of polydopamine/boric acid (PDA/BA). Depending on the selected coating configuration, the concentration of BA in the bulk is critical for favorable crosslinking, where the crosslinking between CNF and BA is demonstrated through a reduction in free OH groups for coatings with the best mechanical performance.

From the experimental evaluation of mechanical coating resistance, a balance between cohesive and adhesive strength has been determined by changing the BA crosslinker concentration in the glass/PDA/BA interface and bulk of the coating. First, an appropriate interface was developed with high BA concentrations required to enhance crosslinking between CNF and PDA/BA while preventing degradation of the PDA. Second, the concentration of the BA crosslinker in the bulk was modulated, resulting in multilayer coatings (CNF/ultra-high BA) with a maximum scratch resistance of 8 N, and mixed-layer coatings (CNF + high BA) with a maximum scratch resistance of 9 N.

The improvement in mechanical resistance of mixed-layer coatings (CNF + BA) is reflected in the higher density of coatings after crosslinking with BA, while the formation of single crystallites in CNF/BA multilayer coatings reflects inferior mechanical properties and brittleness. The crosslinking of mixed-layer coatings with BA is also confirmed by a retarded penetration of water and changes in transparency as a function of the coating density. The chemical interactions through hydrogen bonding after crosslinking of CNF and BA are qualitatively demonstrated, depending on the drying conditions and appropriate concentrations of BA crosslinker in the bulk and interface layer.

As presented in this work, the mechanical robustness of sprayed nanocellulose coatings can be ensured by crosslinking of multilayer and mixed-layer configurations, while the scalability of the spraying process enables its future application in more demanding industrial applications.

Supplementary Materials: The following supporting information can be downloaded at https://www.mdpi.com/article/10.3390/polym17182451/s1: Figure S1: SEM evaluation of dried CNF Valida® fine grade S191C; Figure S2: Mechanical resistance testing for multilayer coating (CNF/BA) and mixed-layer coating (CNF + BA) configurations deposited under "wet" conditions, i.e., without intermediate drying in between the different spraying layers: (a) multilayer coatings, (b) mixed-layer coatings. Only coatings with 5 and 10 sprayed layers are included because of inferior properties of coatings with only 2 sprayed layers; Figure S3: Mechanical resistance testing for mixed-layer coatings with a sprayed PDA layer without a BA crosslinker in the interface, and a sprayed PDA layer with a BA crosslinker in the interface; Figure S4: Detailed microscopic evaluation of crystallite structures for multilayer coatings (CNF/BA) deposited under wet conditions (without intermediate drying) or dry conditions (with intermediate drying); Figure S5: Coating deposition characteristics as a function of the number of sprayed layers, and final characteristics including a coating weight and coating density summary for different coating configurations with 3, 5, and 10 sprayed layers: extended graph with additional data for the sprayed adhesion layer; Figure S6: Reference measurements for water contact angles as a function of time for glass substrates, glass/PDA substrates, and glass/PDA/BA substrates; Figure S7: Illustration of raw data for transparency evaluation, including optical microscopy (left), and UV/VIS spectroscopy (right); Figure S8: Transparency of crosslinked coatings as determined from UV/VIS spectroscopy: extended graph with additional data for sprayed adhesion layer; Figure S9: Infrared spectroscopy for multilayer coatings (CNF/BA) with high BA and ultra-high BA concentrations.

Author Contributions: Conceptualization, G.A.C.; methodology, G.A.C. and P.S.; validation, G.A.C. and P.S.; formal analysis, P.S.; investigation, G.A.C. and P.S.; resources, E.V.V.d.E. and P.C.; data curation, P.S.; writing—original draft preparation, P.S.; writing—review and editing, G.A.C. and E.V.V.d.E.; visualization, G.A.C.; supervision, E.V.V.d.E.; project administration, P.C.; funding acquisition, E.V.V.d.E. and P.C. All authors have read and agreed to the published version of the manuscript.

Funding: This research was funded by the European Union through the Horizon Europe (HORIZON) program under grant number 101057992. Views and opinions expressed are, however, those of the author(s) only and do not necessarily reflect those of the European Union. Neither the European Union nor the granting authority can be held responsible for them.

Institutional Review Board Statement: Not applicable.

Data Availability Statement: The original contributions presented in this study are included in the article/Supplementary Material. Further inquiries can be directed to the corresponding author.

Acknowledgments: The authors acknowledge the technical support of Patrick Doumen and Dylan Broeckaert for sample preparation. This publication has been prepared with the support of the "RUDN University Strategic Academic Leadership Program" (recipient E.V.V.d.E).

Conflicts of Interest: The authors declare no conflicts of interest.

References

1. Shanmugam, K.; Chandrasekar, N.; Balaji, R. Nanocellulose as a sustainable nanomaterial for films and coating layers via spray-coating and applications. In *Interaction of Nanomaterials with Living Cells*; Sheikh, F.A., Majeed, S., Beigh, M.A., Eds.; Springer: Singapore, 2023; pp. 485–556.
2. Lengowski, E.C.; Bonfatti Júnior, E.A.; Coelho Simon, L.; Bolzon de Muniz, G.I.; Sulato de Andrade, A.; Neves Leite, A.; Souza de Miranda Leite, E.L. Nanocellulose coating on kraft paper. *Coatings* **2023**, *13*, 1705. [CrossRef]
3. Tozluoglu, A.; Ates, S.; Durmaz, E.; Setkaya, S.; Arslan, R.; Ozcelik, O.; Candan, Z. Nanocellulose in paper and board coating. In *Emerging Nanomaterials*; Taghiyari, H.R., Morrell, J.J., Husen, A., Eds.; Springer: Cham, Germany, 2022; pp. 197–298.

4. Qi, S.; Kiratzis, I.; Adoni, P.; Tuekprakhon, A.; Hill, H.J.; Stamataki, Z.; Nabi, A.; Waugh, D.; Rodriguez, J.R.; Clarke, S.M.; et al. Porous cellulose thin films as sustainable and effective antimicrobial surface coatings. *ACS Appl. Mater. Interf.* **2023**, *15*, 20638–20648. [CrossRef] [PubMed]
5. Kaur, P.; Sharma, N.; Managala, M.; Rajkhowa, R.; Allardyce, B.; Shastri, Y.; Agrawal, R. Nanocellulose: Resources, physio-chemical properties, current uses and future applications. *Front. Nanotechnol.* **2021**, *3*, 747329. [CrossRef]
6. Usov, I.; Nyström, G.; Adamcik, J.; Handschin, S.; Schütz, C.; Fall, A.; Bergström, L.; Mezzenga, R. Understanding nanocellulose chirality and structure—Properties relationship at the single fibril level. *Nature Commun.* **2015**, *6*, 7564. [CrossRef]
7. Nishiyama, Y. Molecular interactions in nanocellulose assembly. *Phil. Trans. R. Soc. A* **2017**, *376*, 20170027. [CrossRef] [PubMed]
8. Shanmugam, K. Spray-coated cellulose nanofiber films: Preparation, characterization and application. In *Nanocellulose—Sources, Preparations and Applications*; Kazi, S.M., Ed.; Intechopen: London, UK, 2024; pp. 130–179.
9. Pacaphol, K.; Aht-Ong, D. The influences of silanes on interfacial adhesion and surface properties of nanocellulose film coating on glass and aluminum substrates. *Surf. Coat. Technol.* **2017**, *320*, 70–81. [CrossRef]
10. Spagnuolo, L.; D'Orsi, R.; Operamolla, A. Nanocellulose for paper and textile coating: The importance of surface chemistry. *ChemPlusChem* **2022**, *87*, e202200204. [CrossRef]
11. Farooq, A.; Memon, H.; Farooq, A.; Wang, Z. (Eds.) Nanocellulose in textile applications. In *Deep Eutectic Solvents in the Textile Industry*; Springer: Singapore, 2024; pp. 75–88.
12. Saremi, R.; Borodinov, N.; Laradji, A.M.; Sharma, S.; Luzinov, I.; Minko, S. Adhesion and stability of nanocellulose coatings on flat polymer films and textiles. *Molecules* **2020**, *25*, 3238. [CrossRef]
13. Liang, L.; Bhagia, S.; Li, M.; Huang, C.; Ragauskas, A.J. Cross-linked nanocellulosic materials and their applications. *ChemSusChem* **2020**, *13*, 78–87. [CrossRef]
14. Xu, H.; Liu, Y.; Xie, Y.; Zhu, E.; Shoi, Z.; Yang, Q.; Xiong, C. Doubly cross-linked nanocellulose hydrogels with excellent mechanical properties. *Cellulose* **2019**, *26*, 8645–8654. [CrossRef]
15. Hossain, L.; Raghuwanshi, V.S.; Tanner, J.; Wu, C.M.; Kleinerman, O.; Cohen, Y.; Garnier, G. Structure and swelling of cross-linked nanocellulose foams. *J. Coll. Interf. Sci.* **2020**, *568*, 234–244. [CrossRef]
16. Zulfiana, D.; Karimah, A.; Anita, S.H.; Masruchin, N.; Wijaya, K.; Suryanegara, L.; Fatriasari, W.; Fudholi, A. Antimicrobial Imperata cylindrica paper coated with anionic nanocellulose crosslinked with cationic ions. *Int. J. Biol. Macromol.* **2020**, *164*, 892–901. [CrossRef] [PubMed]
17. Xu, J.; Deng, X.; Dong, Y.; Zhou, Z.; Zhang, Y.; Yu, J.; Cai, J.; Zhang, Y. High-strength, transparent and superhydrophobic nanocellulose/nanochitin membranes fabricated via crosslinking of nanofibers and coating F-SiO_2 suspensions. *Carbohydr. Polym.* **2020**, *237*, 116694. [CrossRef]
18. Jeon, H.; Son, J.H.; Lee, J.; Park, S.B.; Ju, S.; Oh, D.X.; Koo, J.M.; Park, J. Preparation of a nanocellulose/nanochitin coating on a poly(lactic acid) film for improved hydrolysis resistance. *Int. J. Biol. Macromol.* **2024**, *254*, 127790. [CrossRef] [PubMed]
19. Garg, R.; Rana, H.; Singh, N.; Goswami, S. Guargum/nanocellulose based novel crosslinked antimicrobial film with enhanced barrier and mechanical properties for food packaging. *J. Environ. Chem. Eng.* **2023**, *11*, 109254. [CrossRef]
20. Dai, L.; Long, Z.; Chen, J.; An, X.; Cheng, D.; Khan, A.; Ni, Y. Robust guar gum/cellulose nanofibrils multilayer films with good barrier properties. *ACS Appl. Mater. Interf.* **2017**, *9*, 5477–5485. [CrossRef] [PubMed]
21. Matsumura, H.; Sugiyama, J.; Wolfgang, G. Cellulosic nanocomposites. I. Thermally deformable cellulosic hexanoates from heterogeneous reaction. *J. Appl. Polym. Sci.* **2000**, *78*, 2242–2253. [CrossRef]
22. Vuoti, S.; Talja, R.; Johansson, L.; Heikkinen, H.; Tammelin, T. Solvent impact on esterification and film formation ability of nanofibrillated cellulose. *Cellulose* **2013**, *20*, 2359–2370. [CrossRef]
23. De Cuadro, P.; Belt, T.; Kontturi, K.; Kontturi, E.; Vuorinen, T.; Hughes, M. Cross-linking of cellulose and poly(-ethylene glycol) with citric acid. *React. Func. Polym.* **2015**, *90*, 21–24. [CrossRef]
24. Zhou, Y.J.; Luner, P.; Caluwe, P. Mechanism of cross-linking of paper with polyfunctional carboxylic acids. *J. Appl. Polym. Sci.* **1995**, *58*, 1523–1534. [CrossRef]
25. Quellmalz, A.; Mihranyan, A. Citric acid cross-linked nanocellulose-based paper for size-exclusion nanofiltration. *ACS Biomater. Sci. Eng.* **2015**, *1*, 271–276. [CrossRef]
26. Sirvio, J.; Kolehmainen, A.; Liimatainen, H.; Niinimaki, J.; Hormi, O. Biocomposite cellulose-alginate films: Promising packaging materials. *Food Chem.* **2014**, *151*, 343–351. [CrossRef]
27. Csiszár, E.; Herceg, I.; Fekete, E. Effect of heating and citric acid on the performance of cellulose nanocrystal thin films. *Polymers* **2023**, *15*, 1698. [CrossRef] [PubMed]
28. Yang, H.; Edberg, J.; Gueskine, V.; Vagin, M.; Say, M.G.; Erlandsson, J.; Wagberg, L.; Engquist, I.; Berggren, M. The effect of crosslinking on ion transport in nanocellulose-based membranes. *Carbohydr. Polym.* **2022**, *278*, 118938. [CrossRef]
29. Cherian, R.M.; Tharayil, A.; Varghese, R.T.; Antony, T.; Kargarzadeh, H.; Chirayil, C.J.; Thomas, S. A review on the emerging applications of nano-cellulose as advanced coatings. *Carbohydr. Polym.* **2022**, *282*, 119123. [CrossRef]

30. Hossain, L.; Ledesma, R.M.; Tanner, J.; Garnier, G. Effect of crosslinking on nanocellulose superabsorbent biodegradability. *Carbohydr. Polym. Technol. Appl.* **2022**, *3*, 100199. [CrossRef]
31. Hubbe, M.A.; Tayeb, P.; Joyce, M.; Tyagi, P.; Kehoe, M.; Dimic-Misic, K.; Pal, L. Rheology of nanocellulose-rich aqueous suspensions: A review. *BioResources* **2017**, *12*, 9556–9661. [CrossRef]
32. Thakur, V.; Guleria, A.; Kumar, S.; Sharma, S.; Singh, K. Recent advances in nanocellulose processing, functionalization and applications: A review. *Mater. Adv.* **2021**, *2*, 1872–1895. [CrossRef]
33. Hossain, L.; Raghuwanshi, V.S.; Tanner, J.; Garnier, G. Modulating nanocellulose hydrogels and cryogels strength by crosslinking and blending. *Coll. Surf. A* **2021**, *630*, 127608. [CrossRef]
34. Herrera, M.A.; Mathew, A.P.; Oksman, K. Barrier and mechanical properties of plasticized and cross-linked nanocellulose coatings for paper packaging applications. *Cellulose* **2017**, *24*, 30969–33980. [CrossRef]
35. Zhang, Z.; Li, Y.; Song, L.; Ren, L.; Xu, X.; Lu, S. Swelling resistance and water-induced shape memory performances of sisal cellulose nanofibers/polyethylene glycol/citric acid nanocellulose papers. *J. Nanomater.* **2019**, *1*, 4304532. [CrossRef]
36. Hadrup, H.; Fredriksen, M.; Sharma, A.K. Toxicity of boric acid, borax and other boron containing compounds: A review. *Reg. Tox. Pharm.* **2021**, *121*, 104873. [CrossRef]
37. Li, K.; Jin, S.; Wei, Y.; Li, X.; Li, J.; Shi, S.Q.; Li, J. Bioinspired hyperbranched protein adhesive based on boronic acid-functionalized cellulose nanofibril and water-soluble polyester. *Comp. B* **2021**, *219*, 108943. [CrossRef]
38. Stubelius, A.; Lee, S.; Almutairi, A. The chemistry of boronic acids in nanomaterials for drug delivery. *Acc. Chem. Res.* **2019**, *52*, 3108–3119. [CrossRef]
39. Van Duin, M.; Peters, J.A.; Kieboom, A.P.G.; Van Bekkum, H. Studies on borate esters: The pH dependence of the stability of esters of boric acid and borate in aqueous medium as studied by 11B NMR. *Tetrahedron* **1984**, *40*, 2901–2911. [CrossRef]
40. Hall, D.G. (Ed.) Structure, properties, and preparation of boronic acid derivatives. Overview of their reactions and applications. In *Boronic Acids: Preparation and Applications in Organic Synthesis and Medicine*; Wiley-VCH Verlag: Weinheim, Germany, 2005; pp. 1–99.
41. Rietjens, M.; Steenbergen, P.A. Crosslinking mechanism of boric acid with diols revisited. *EurJIC* **2005**, *6*, 1162–1174. [CrossRef]
42. Ishihara, K.; Mouri, Y.; Funahashi, S.; Tanaka, M. Mechanistic study of the complex formation of boric acid. *Inorg. Chem.* **1991**, *30*, 2356–2360. [CrossRef]
43. Raval, P.; Thomas, N.; Hamdouna, L.; Delevoye, L.; Lafon, O.; Reddy, G.N. Boron adsorption kinetics of microcrystalline cellulose and polymer resin. *Langmuir* **2023**, *39*, 5384–5395. [CrossRef]
44. Shen, H.; Zheng, X.; Dong, L.; Huang, D. Bioinspired high-strength borate cross-linked microfibrillated cellulose composite laminate with self-extinguishing flame retardance and superhydrophobicity for self-cleaning. *ACS Omega* **2023**, *8*, 41458–41468. [CrossRef]
45. Yildirim, M.; Candan, Z.; Gonultas, O. Chemical performance analysis of nanocellulose/boron-compound-reinforced hybrid UF resin. *Green Mater.* **2022**, *10*, 90–96. [CrossRef]
46. Yildirim, M.; Candan, Z. Performance properties of particleboard panels modified with nanocellulose/boric acid. *BioResources* **2021**, *16*, 1875–1890. [CrossRef]
47. Moreno, P.; Villamizar, N.; Perez, J.; Bayona, A.; Roman, J.; Moreno, N.; Cardozo, N.S.M. Fire-resistant cellulose boards from waste newspaper, boric acid salts, and protein binders. *Clean Technol. Environ.* **2021**, *23*, 1537–1546. [CrossRef]
48. Wicklein, B.; Kocjan, D.; Carosio, F.; Camino, G.; Bergström, L. Tuning the nanocellulose-borate interaction to achieve highly flame retardant hybrid materials. *Chem. Mater.* **2016**, *28*, 1985–1989. [CrossRef]
49. Fu, D.; Yang, R.; Wang, R.; Wang, Y.; Li, Y.; Bian, H. A triple-crosslinked, self-healing, polyvinyl alcohol/nanocellulose hydrogel for versatile sensing application. *Carbohydr. Polym.* **2024**, *334*, 122060. [CrossRef]
50. Zhang, L.; Wan, C.; Su, J.; Zhang, W.; Wei, S.; Tian, W.; Liu, X.; Cheng, W.; Li, X.; Li, X.; et al. A dual-crosslinked self-healing and antibacterial nanocellulose hydrogel for monitoring of human motions. *Mater. Des.* **2022**, *215*, 110464. [CrossRef]
51. Spoljaric, S.; Salminen, A.; Luong, N.D.; Seppala, J. Stable, self-healing hydrogels from nanofibrillated cellulose, poly(vinyl alcohol) and borax via reversible crosslinking. *Eur. Polym. J.* **2014**, *56*, 105–117. [CrossRef]
52. Gadhave, R.V.; Vineeth, S.K.; Mahanwar, P.A.; Gadekar, P.T. Effect of addition of boric acid on thermo-mechanical properties of microcrystalline cellulose/polyvinyl alcohol blend and applicability as wood adhesive. *J. Adhes. Sci. Technol.* **2021**, *35*, 1072–1086. [CrossRef]
53. Rouhi, M.; Garavand, F.; Heydari, M.; Mohammadi, R.; Sarlak, Z.; Cacciotti, I.; Razavi, S.H.; Mousavi, M.; Parandi, E. Fabrication of novel antimicrobial nanocomposite films based on polyvinyl alcohol, bacterial cellulose nanocrystals, and boric acid for food packaging. *J. Food Meas. Char.* **2024**, *18*, 2146–2161. [CrossRef]
54. Uddin, K.M.A.; Ago, M.; Rojas, O.J. Hybrid films of chitosan, cellulose nanofibrils and boric acid: Flame retardancy, optical and thermo-mechanical properties. *Carbohydr. Polym.* **2017**, *177*, 13–21. [CrossRef]
55. Yaylaci, E.U. Antibacterial Effects of Boric Acid Against Aquatic Pathogens. *J. Anat. Eviron. Anim. Sci.* **2021**, *6*, 240–244.

56. Durmaz, E.; Ates, S. Characterisation of surfaces coated with different nanocellulose-based suspensions. *Int. J. Surf. Sci. Eng.* **2023**, *17*, 141–164. [CrossRef]
57. Feinberg, H.; Hanks, T.W. Polydopamine: A bioinspired adhesive and surface modification platform. *Polym. Internat.* **2021**, *71*, 578–582. [CrossRef]
58. Lee, H.A.; Ma, Y.; Zhou, F.; Hong, S.; Lee, H. Material-independent surface chemistry beyond polydopamine coating. *Acc. Chem. Res.* **2019**, *52*, 704–713. [CrossRef] [PubMed]
59. Yang, F.K.; Zhao, B. Adhesion properties of self-polymerized dopamine thin film. *Open Surf. Sci. J.* **2011**, *3*, 115–122. [CrossRef]
60. Samyn, P. Polydopamine and cellulose: Two biomaterials with excellent compatibility and applicability. *Polym. Rev.* **2021**, *61*, 814–865. [CrossRef]
61. Liu, Y.; Sui, Y.; Liu, C.; Liu, C.; Wu, M.; Li, B.; Li, Y. A physically crosslinked polydopamine/nanocellulose hydrogel as potential versatile vehicles for drug delivery and wound healing. *Carbohydr. Polym.* **2018**, *188*, 27–36. [CrossRef] [PubMed]
62. Petersson, F.; Jonsson, M. The reactivity of hydroxyl radicals toward boric acid as a function of pH. *J. Phys. Chem. A* **2024**, *128*, 7593–7600. [CrossRef]
63. Lee, H.; Dellatore, S.M.; Miller, W.M.; Messersmith, P.B. Mussel-inspired surface chemistry for multifunctional coatings. *Science* **2007**, *318*, 5849. [CrossRef]
64. Hong, S.H.; Hong, S.; Ryou, M.H.; Choi, J.W.; Kang, S.M.; Lee, H. Sprayable ultrafast polydopamine surface modifications. *Adv. Mater. Interfac.* **2016**, *3*, 1500857. [CrossRef]
65. Graveson, I.; English, R. Low Energy Method for the Preparation of Non-Derivatized Nanocellulose. U.S. Patent US20150158955A1, 11 June 2015.
66. *ISO 1518-1*; Paints and Varnishes—Determination of Scratch Resistance Part 1: Constant-Loading Method. International Organization for Standardization: Geneva, Switzerland, 2023.
67. *ISO 2409*; Paints and Varnishes—Cross-Cut Test. International Organization for Standardization: Geneva, Switzerland, 2020.
68. *ISO 25178-2*; Geometrical Product Specifications (GPS)—Surface Texture: Areal, Part 2: Terms, Definitions and Surface Texture Parameters. International Organization for Standardization: Geneva, Switzerland, 2021.
69. *ISO 19403-2*; Paints and Varnishes—Wettability, Part 2: Determination of the Surface Free Energy of Solid Surfaces by Measuring the Contact Angle. International Organization for Standardization: Geneva, Switzerland, 2024.
70. Samyn, P.; Cosemans, P.; Chandroth, A.M.; Takahashi, K.; Everaerts, J. Mechanical properties of spray-coated nanocellulose coatings. *Surf. Coat. Technol.* **2025**, *496*, 131601. [CrossRef]
71. Schneider, A.; Hemmerlé, J.; Allais, M.; Didierjean, J.; Michel, M.; d'Ischia, M.; Ball, V. Boric acid as an efficient agent for the control of polydopamine self-assembly and surface properties. *ACS Appl. Mater. Interf.* **2018**, *20*, 7574–7580. [CrossRef]
72. Alavia, W.; Seidel-Morgenstern, A.; Hermsdorf, D.; Lorenz, H.; Graber, T.A. Real-time crystal growth monitoring of boric acid from sodium or lithium sulfate containing aqueous solutions by atomic force microscopy. *ACS Omega* **2023**, *8*, 10822–10835. [CrossRef]
73. Gao, Y.; Radaniello, W.; Toksoy, M.F.; Munhollon, T.; Haber, R. Improvement of crystallization and particle size distribution of boric acid in the processing of a boron carbide precursor. *RCS Adv.* **2015**, *25*, 19067–19073. [CrossRef]
74. Harabor, A.; Rotaru, P.; Scorei, R.I.; Harabor, N.A. Non-conventional hexagonal structure for boric acid. *J. Therm. Anal. Calor.* **2014**, *118*, 1375–1384. [CrossRef]
75. Jean, J.H. Crystallization of boric acid on low-k glass+ ceramic green tapes. *Jpn. J. Appl. Phys.* **1996**, *35*, L429. [CrossRef]
76. Pan, R.; Cheng, Y.; Pei, Y.; Liu, J.; Tian, W.; Jiang, Y.; Tang, K.; Zhang, J.; Zheng, X. Cellulose materials with high light transmittance and high haze: A review. *Cellulose* **2023**, *30*, 4813–4826. [CrossRef]
77. Ansari, F.; Ding, Y.; Berglund, L.A.; Dauskardt, R.H. Toward sustainable multifunctional coatings containing nanocellulose in a hybrid glass matrix. *ACS Nano* **2018**, *12*, 5495–5503. [CrossRef] [PubMed]
78. Han, J.; Yue, Y.; Wu, Q.; Huang, C.; Pan, H.; Zhan, X.; Mei, C.; Xu, X. Effects of nanocellulose on the structure and properties of poly(vinyl alcohol)-borax hybrid foams. *Cellulose* **2017**, *24*, 4433–4448. [CrossRef]
79. Zhang, J.; Koubaa, A.; Xing, D.; Liu, W.; Wang, Q.; Wang, X.M.; Wang, H. Improving lignocellulose thermal stability by chemical modification with boric acid for incorporating into polyamide. *Mater. Des.* **2020**, *191*, 108589. [CrossRef]
80. Durmaz, E.; Ates, S. Effect of nanocellulose type and matrix material on production of nanocomposites. *Cell. Chem. Technol.* **2023**, *57*, 625–635. [CrossRef]

polymers

MDPI

Article

Agricultural Waste-Derived Cellulose/ZnO Composites: Dual Photocatalytic and Adsorptive Action for Textile Dye Removal

Jihene Belhaj [1,2], Ramzi Khiari [3,4], Valentín García-Caballero [5], Antonio A. Romero [1,6] and Araceli García [1,6,*]

1 Nanoval FQM-383 Research Group, Organic Chemistry Department, University of Córdoba, Marie Curie (C-3) Building, Crta. Nnal. km 396, E-14014 Córdoba, Spain; z12bebej@uco.es (J.B.); qo1rorea@uco.es (A.A.R.)
2 National Engineering School of Monastir, University of Monastir, Monastir 5019, Tunisia
3 Department of Textile, Higher Institute of Technological Studies of Ksar Hellal, Ksar Hellal 5070, Tunisia; khiari_ramzi2000@yahoo.fr
4 CNRS, Grenoble INP, LGP2, Université Grenoble Alpes, F-38000 Grenoble, France
5 Departamento de Química Física y Termodinámica Aplicada, Instituto Químico para la Energía y el Medioambiente (IQUEMA), University of Córdoba, E-14014 Córdoba, Spain; g32gacav@uco.es
6 Instituto Químico Para la Energía y el Medioambiente (IQUEMA), Faculty of Sciences, University of Córdoba, Marie Curie (C-3) Building, Crta. Nnal. km 396, E-14014 Córdoba, Spain
* Correspondence: qo2ganua@uco.es

Abstract

The synthesis of cellulose extracted from agricultural waste, specifically almond and fig tree trimmings, and its combination with ZnO nanoparticles to form cellulose/ZnO composites was studied. These adsorbents/photocatalysts were fully characterized, confirming not only the effective deposition of zinc oxide nanoparticles on the cellulose surface but also the improvement in homogeneity and lower agglomeration and size of ZnO particles grown on these fibers (crystallites were 43 ± 12 nm for pristine ZnO and 13–26 nm for composites). The efficacy of these composites was evaluated against methylene blue (MB), methyl orange (MO), and bromophenol blue (BB), this study being the first time that BB removal results have been reported using dual photo-adsorptive cellulosic composites. After 20 min, removals of approximately 45% were achieved for the anionic dyes MO and BB under UV light and up to 65% for MB with either applied radiation, indicating a clear adsorption mechanism for this cationic dye. A reusability study was conducted for the BB removal system, with only a 15–19% loss in BB removal capacity under UV irradiation after the third reuse. These results demonstrated the potential and efficiency of cellulose/ZnO composites as promising photocatalysts for textile wastewater treatment, providing a sustainable and interesting approach to mitigate dye pollution.

Keywords: agriculture waste; photocatalytic degradation; dye adsorption; cellulose/ZnO composites; bromophenol blue

1. Introduction

The global textile industry plays a key role in the global economy. As of 2024, its market size is estimated at approximately USD 1.7 trillion, with projections suggesting growth to more than USD 3 trillion by 2030. However, the fashion sector remains the second highest industrial contributor to waste generation and pollution, accounting for 8% of global carbon emissions and 20% of all wastewater [1]. The textile industry consumes approximately 93 billion cubic meters of water, equivalent to the annual consumption of 5 million people. This large water usage is primarily caused by dyeing and finishing processes, which require substantial amounts of water for fabric treatment and coloration. Over the past two decades, significant efforts have been made to address the environmental impact of dyes in industrial wastewater [1–3] due to the large volumes generated and the vast array of compounds involved. More than 100,000 types of dyes are commercially available, with annual production exceeding 700,000 tons. Due to their high solubility, synthetic dyes are common water pollutants, often present in trace amounts in industrial wastewater [4]. As organic content regulations become more stringent, the removal of dyes before discharge is crucial. Many of these dyes are toxic and pose serious risks to water quality, plant and animal life, and human health, for which prolonged exposure has been linked to skin and eye irritation, corneal and conjunctival lesions, dermatitis, and even cancer [5].

In recent years, there has been a strong research trend in the development of new biomass-derived materials as photocatalysts for the treatment of this textile-contaminated industrial water [6–8]. One of these promising materials is cellulose, a biodegradable and renewable polymer derived from agricultural and food waste that has been extensively studied for its potential in various applications, including water purification [9] or energy storage [10]. As an improvement, researchers have explored the synthesis of cellulose-based composites with other materials, such as metal oxides [11–14] in a symbiosis that combines the best of a biodegradable and sustainably sourced matrix with the proven dye-removing efficacy [4,15,16] and the antibacterial activities of metal oxides [17–19], with even more effective properties achieved by combining the two through the development of cellulose and metal oxide composites [20].

ZnO, together with TiO_2 particles, is one of the semiconductors most extensively studied as photocatalysts [21,22], gaining considerable attention for its low production cost, biocompatibility, narrow bandgap energy, flame retardancy capacity, and other beneficial properties [23,24]. In addition to its catalytic properties, ZnO has been used in the adsorption of organic dyes [25] because due to the ionization of Zn−OH groups on its surface, this material has a point of zero charge (PZC) between 9 and 11 and favors electrostatic attraction and hydrogen bond attraction, π–π interactions, and Van der Waals forces [24], improving the adsorption of anionic molecules (such as methyl orange, bromophenol blue, or amaranth) in acidic media or cationic molecules (methylene blue) under basic conditions. Although ZnO is capable of adsorbing dyes, the better the greater the porosity and specific surface of the material, its greatest advantage is that it acts as a photocatalyst, degrading dyes under UV or visible irradiation. In these processes, the dye is first adsorbed and then degraded by reactive species generated by ZnO.

However, the efficiency and applicability of ZnO is limited by its agglomeration tendency [26], which significantly affects its photocatalytic performance [23,27]. One solution involves the deposition of ZnO particles on a polymeric support or substrate [28], which provides an active surface that allows for a homogeneous dispersion of the nucleation sites and prevents particle accumulation [23,29] in addition to improving its photocatalytic properties, such as a broadened absorption edge and a wider wavelength range to achieve pollutant removal. This solution, using different types of cellulose (different size, functionality, plant or bacterial origin…) has become of great interest in last few years [21,30], with researchers seeking a direct application in the removal of coloring molecules used in the textile industry, such as crystal violet, methylene blue, methyl orange, or other aqueous contaminants such as phenols or pharmaceuticals.

To the best of our knowledge, there are no published works evaluating the removal of these three different textile dyes using cellulose composites with zinc oxide. The present study focuses on the synthesis of cellulose/ZnO composites derived from agri-food wastes for the removal of textile dyes by an alkaline deposition method using zinc acetate dihydrate as a precursor. The different materials were analyzed to determine their structure, morphology, and chemical functionality (surface charge, FT-IR, XRD, XPS, and porosimetry). Furthermore, the optical properties (bandgap) of the obtained cellulose/ZnO composites were evaluated and compared with those of ZnO particles prepared by the same method. Finally, their ability to remove dyes was evaluated, studying the two possible mechanisms that can occur and their potential synergy: adsorption (tests conducted in the dark) and photodegradation (experiments irradiated with UV and white light) as well as how the properties of the materials evaluated influence them. The use of agricultural and food waste-derived cellulose as a precursor for the synthesis of these composites offers a sustainable and environmentally friendly approach to the development of less sustainable or more harmful materials.

2. Materials and Methods

2.1. Materials

Two agricultural residues were used as cellulose source, namely almond (*Prunus amygdalus* L.) and fig (*Ficus carica* L.) trimmings. The raw materials were collected in January 2022 in Ksour Essaf, Mahdia, Tunisia. After harvesting, the agricultural wastes were dried under ambient conditions (65% humidity, 25 °C) until constant weight, after which sand was removed by conventional washing and finally dried again under the previous conditions prior to their use.

For the preparation of cellulose and cellulosic composites, sodium hydroxide pellets (≥99.0%, AGR, from Labkem), hydrogen peroxide (30%, GLR, from Labkem), and zinc acetate dihydrate (≥98%, ACS, from Labkem) were used as supplied. Deionized water was used for washing and sample preparation.

The performance of cellulosic composites in water decontamination processes was evaluated using methyl orange (MO, C.I. 13025 ACS), methylene blue (MB, C.I. 52015 IND), and bromophenol blue (BB, IND ACS), all supplied by Labkem (Barcelona, Spain) and used as received.

2.2. Cellulose Extraction

The raw materials were first chipped into uniform pieces in an automatic jaw crusher with 0.5 cm sieve (Retsch GmbH, Haan, Germany). Then, a pressurized and thermostatized 15 L reactor (Metrotec Techlab Systems, Lezo, Spain) was used to delignify the chips with an 8:1 liquid/solid ratio and 10% NaOH at 170 °C for 60 min. The treated material was subsequently thoroughly washed, disintegrated in a PFI grinder (Metrotec Techlab Systems, Lezo, Spain) at 1200 rpm for 30 min, filtered through a 50 μm mesh, and oven dried (105 ± 3 °C) until constant weight [31]. The resulting pulp was bleached to remove impurities such as proteins, phenolic compounds, and remaining lignin. The bleaching process started with 30 g of dry fiber mixed with 200 mL of 5% H_2O_2 aqueous solution and 2–3 mL of 15% NaOH to raise the pH and was conducted for 120 min at 80–85 °C, followed by washing and filtration. This step was repeated 2–3 times until complete bleaching. The bleached fibers obtained from almond tree and fig tree trimmings, hereinafter abbreviated as ATcell and FTcell, respectively, were then oven dried at 105 ± 3 °C until constant weight and hermetically stored for later use.

The morphology of ATcell and FTcell fibers was assessed using a MORFI (LB-01) analyzer, developed by Techpap (Gières, France). This analysis equipment uses an image analysis technique to determine the main morphological parameters of a fiber suspension in terms of the arithmetic mean of fiber length (L, mm) and fiber width (W, μm). The measurements were conducted in triplicate with 1 L of aqueous of 0.03% wt. of cellulosic pulp [32].

2.3. Cellulose/ZnO Composite Preparation

Cellulose/ZnO composites were prepared by an in situ solution casting technique, using 2.2 g of previously prepared cellulose fibers dispersed in 100 mL of 0.5 mol·L^{-1} zinc acetate dihydrate aqueous solution. The mixture was heated at 90 °C for 60 min under continuous magnetic stirring [19]. Subsequently, 0.5 mol·L^{-1} NaOH was dropwise added to the solution while maintaining the same conditions (90 °C and the magnetic stirring) until a precipitate was formed, then recovered by vacuum filtration (nylon membrane filters of 0.45 μm, Ø 47 mm, Labbox, Barcelona, Spain) and thoroughly washed with deionized water and oven dried (100 ± 3 °C for 24 h), ensuring the complete formation of ZnO without intermediate compounds [33]. This procedure was applied for ATcell and FTcell samples, obtaining the corresponding ATcell/ZnO and FTcell/ZnO cellulosic composites. The inorganic matter (IM) of the prepared celluloses and cellulosic composites, which should be considered entirely due to the zinc oxide deposited on cellulose surface, was determined by gravimetry after combustion of each sample in a ceramic crucible at 575 °C for 4 h in a laboratory furnace [34]. The IM was calculated according to Equation (1).

$$IM(\%) = \frac{m_f - m_i}{m_s} \times 100 \quad (1)$$

where m_f and m_i are the mass (g) of dry crucible with sample after and before burning and m_s is the mass of sample. This measurement was performed in duplicate for each sample.

2.4. Sample Characterization

For detailed information about performed analytical techniques, please refer to Supplementary Information (SI) Text S1 [35,36].

2.5. Removal of Dyes

Experiments were conducted in triplicate with 0.1 g of each cellulosic sample suspended into 50 mL of 5 ppm aqueous solution of each dye, i.e., methylene blue (MB), bromophenol blue (BB), and methyl orange (MO) [37]. Each experiment was kept for 20 min under constant stirring (200 rpm) in a photoreactor prototype [38], and different light sources/conditions were considered: UV light, white light, and darkness (see Text S2 in SI for more details). Removal of dyes was monitored by collecting aliquots at regular intervals and analyzed using a UV/VIS spectrophotometer Lambda 365 UV-Vis spectrophotometer (PerkinElmer, Waltham, MA, USA) after filtration with syringe nylon filters (0.45 μm).

The absorbance spectra of the supernatants were recorded at wavelengths of 664 nm, 598 nm, and 464 nm for MB, BB, and MO removal experiments, respectively, corresponding to the respective maximum wavelength λ_{max} (see Table S1). Dye removal (R%) was calculated based on the recorded absorbance according to the Equation (2).

$$R(\%) = \frac{(ABS_0 - ABS_t)}{ABS_0} \times 100 \quad (2)$$

where ABS_0 and ABS_t are the absorbances of the supernatants at the beginning and at each time of the experiment, respectively.

2.6. Regeneration and Reuse of Composites

From a sustainability perspective in the treatment of wastewater from the textile industry, the regeneration and recycling of adsorbent and catalytic materials results in more than necessary [29]. After each performed test, the composite was recovered by vacuum filtration (low ash content filter paper, BRANCHIA by PRAT-DUMAS equivalent to Whatman 52), rinsed 4 times with deionized water to remove residual non-adsorbed dye [39], and dried (24 h at 80 °C) for its analysis and subsequent regeneration/reuse study.

The reusability of cellulose/ZnO composites was evaluated for BB removal system by treating the adsorbed catalyst with 5 mL of absolute ethanol [29] followed by centrifugation (3000 rpm, 10 min) and ethanolic solution removal. This was carried out 3 times, after which the material was dried in an oven (24 h, 80 °C) before being reused. The reusability of the composites was conducted in duplicate.

2.7. Statistical Analysis

The dye removal process was carried out in duplicate. The results were expressed as mean values ± standard deviation. The evaluation on the influence of different parameters over dye removal assays was assessed by multiple variable analysis using the software STATGRAPHICS centurion XV, version 15.2.11 (StatPoint, Inc., Warrenton, VA, USA). Pearson correlation parameters and p-values were calculated for each pair of variables, with $p < 0.05$ reflecting significance, in order to assess the linear and significant relationship between them.

3. Results and Discussion

3.1. Characterization of Samples

Cellulose fibers from pruning wastes of almond and fig trees were obtained in a previous work [32]. As shown in Table 1, the morphology of ATcell and FTcell fibers was very similar to each other (0.54–0.57 mm in length and 19.3–22.7 µm in width) and coincided with most fibers obtained from annual plants in conventional biomass fractionation biorefinery processes. Several studies have been reported in the literature indicating that the size of cellulose fibers may be a determining factor in the deposition of ZnO particles on their surface. Li et al. [26] found that morphology affects the chemical deposition of ZnO, yielding higher content with the decrease in fiber size. Similarly, other authors concluded that not only the amount but also ZnO particle size and morphology [5] are directly related and can be controlled by varying the diameter of micro/nano-cellulose fibers with a grinding treatment [40]. Considering the present results on fiber size, no major difference in the ZnO content of the composites prepared from ATcell and FTcell would be expected since both materials have similar sizes.

Table 1. Some chemical, morphological, and structural properties of cellulosic samples (fiber length L, fiber width W, inorganic content IM, surface charge ρ, crystalline index CrI, surface area S_{BET}, and average pore size D_{BJH}).

Sample	L [a] (mm)	W [a] (µm)	IM [b] (%)	ρ [c] (µmol/g)	CrI [d] (%)	S_{BET} [e] (m^2/g)	D_{BJH} [e] (Å)
ATcell	0.57 ± 0.04	19.3 ± 0.6	1.3 ± 0.2	51.0 ± 1.0	66.34	1.80	798
FTcell	0.54 ± 0.03	22.7 ± 0.3	2.0 ± 0.8	79.81 ± 1.7	66.42	3.32	563
ATcell/ZnO	n.d	n.d	47.2 ± 0.8	399.9 ± 10.1	49.68	14.48	332
FTcell/ZnO	n.d	n.d	47.3 ± 0.7	343.7 ± 21.1	52.21	14.94	361
ZnO	n.d	n.d	97.2 ± 0.4	836.2 ± 21.1	n.d	13.42	381

(n.d) not determined for this sample; (a) determined by MORFI analysis; (b) gravimetrically determined according to Equation (1); (c) determined by Equation (S1) after conductometric titration; (d) determined for cellulosic materials by Equation (S2) from XRD analyses; (e) determined by N_2 adsorption/desorption porosimetry.

This behavior is clearly demonstrated by the inorganic (IM in Table 1). Both ATcell and FTcell presented ash contents below 2% wt., attributed to residues of chemicals used during their production. After alkaline treatment with zinc acetate, the samples presented inorganic contents of approximately 47% in both ATcell/ZnO and FTcell/ZnO, which indicated that the growth performance of zinc oxide particles on their surface was almost equal.

The presence of surface functionalities, especially negatively charged ones such as hydroxyl groups, can serve as a zinc oxide nanoparticle growth point and improve their dispersion on the surface of a substrate [15]. Thus, as asserted by Li et al. [40], the uniform nucleation Zn^{2+} to form ZnO particles on cellulosic fibers strongly depends on the preparation pathway and the existence of −OH groups on the cellulose surface. As confirmed by the conductometric titration results in Table 1, ATcell and FTcell celluloses showed certain surface charges of 51.0 and 79.81 μmol/g, respectively, due to hydroxyl and acidic groups, the last probably formed during the samples' preparation [41], confirming the great importance of the origin of the biomass in the behavior of the cellulose extracted from it. For the cellulose/ZnO samples, this charge increased considerably to 340–400 μmol/g, proving again the chemical deposition of ZnO particles (836.2 μmol/g) on the surface of the fibers since these are strongly positively charged [25].

The FT-IR spectra of cellulosic samples are shown in Figure 1. The peaks at 3330 cm^{-1} and 2890 cm^{-1} represent the stretching vibration of hydroxyl and methyl groups in polysaccharides, such as cellulose. Additional absorption bands at 1430, 1370, 1315, 1160, and 1030 cm^{-1} are associated with stretching and bending vibrations of $-CH_2$, −CH, −OH, C−O, and C−O−C bonds in cellulose [39]. The 1430 cm^{-1} band indicates the amount of crystalline structure (less intense after functionalization with ZnO), while the 900 cm^{-1} band represents β-glycosidic linkages between glucose units in cellulose [41].

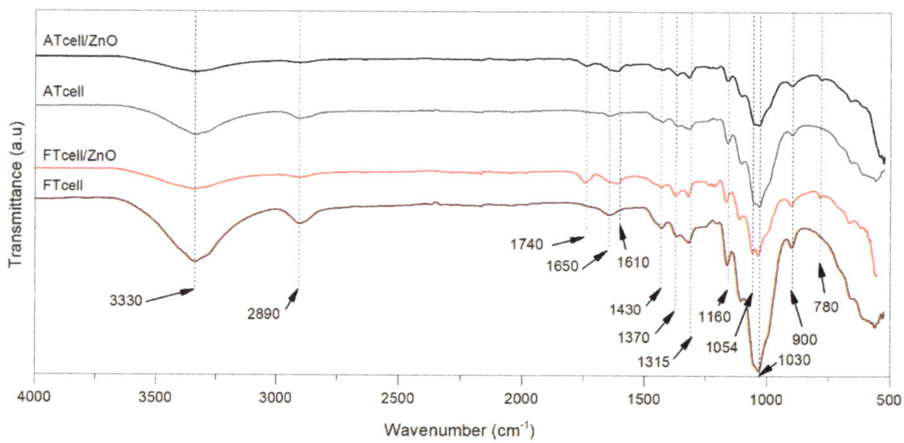

Figure 1. FT-IR for obtained celluloses (ATcell and FTcell) and prepared composites (ATcell/ZnO and FTcell/ZnO).

Although ZnO vibration modes have been reported to appear in the range of 500–400 cm^{-1} [42], some IR bands detected in the present study have already been related to the formation of ZnO particles and their interaction with the cellulose surface [5]. Two bands at 1050 cm^{-1} and 780 cm^{-1} have been assigned in the literature to successful ZnO formation and its interaction with cellulose [13,16], appearing in both cell/ZnO spectra but being more noticeable for FTcell/ZnO. The interaction of ZnO particles with the cellulose structure can also be considered by the intensity reduction in the broad −OH band at 3300 cm^{-1}, suggesting that ZnO particles not only have a strong interaction with the −OH groups of cellulose [5,13,22], but also that the intensity reduction would be due to the growth of the oxide particles directly on the hydroxyl groups available on the cellulose surface. This has been previously proposed in the literature [25,26], thus resulting in the reduction of the band at 3300 cm^{-1} being more evidence of the successful deposition of ZnO on ATcell and FTcell fibers. The significant change on the 2890 cm^{-1} band of cellulose/ZnO composites has been related to the successful functionalization of cellulose fibers with ZnO [5], probably due to hydrogen bond interactions or the formation of cellulose acetate. This last mechanism could be justified by the appearance of a band at 1740 cm^{-1} in both ZnO composites, which has been related in the literature to the stretching vibration of the carbonyl in acetyl groups [14]. In the present work, a slight shift of the band from 1650 cm^{-1} to a lower wavelength in ATcell and FTcell samples was observed in the cellulose/ZnO composite samples, related to the presence of water in the cellulose structure.

Optical microscopy was used to examine the morphology and distribution of pure ZnO particles and cellulose/ZnO composite materials at three different magnifications. As observed in Figure S2 in SI, the ZnO particles had irregular, almost spherical shapes, with a marked tendency to agglomerate due to their high surface energy and Van der Waals interactions. Particle sizes varied from submicrons to a few microns. However, whatever the fiber used, the cellulose/ZnO composites showed a more homogeneous distribution of ZnO particles within the cellulose fiber. The ZnO is largely present on the surface of the fiber. These observations suggest good interfacial compatibility between the two components and good incorporation of the ZnO into the cellulose structure.

The morphological and microstructural changes in the cellulose/ZnO composite modified with ZnO particles were also examined using scanning electron microscopy (SEM), and the results are shown in Figure 2. The images reveal significant alterations in the external surface of the cellulose fibers. Following the saturation of the fibers with ZnO precursors and the in situ synthesis of nanoparticles, a new surface layer is formed. This layer exhibits increased roughness and noticeable nanoparticle agglomeration [5].

To assess the uniformity of the ZnO film deposition, SEM imaging was used to evaluate the surface morphology and particle distribution across the composite. The homogeneity of the ZnO nanoparticle layer was determined based on the consistency of the coating and the absence of large uncoated areas or excessive clustering. Additionally, energy-dispersive X-ray spectroscopy (EDX), coupled with SEM, was employed to confirm the elemental composition and spatial distribution (mapping) of zinc (Zn) and oxygen (O) across the fiber surface. Figure 2 further supports these observations by presenting the ATcell/ZnO composite that exhibits a uniform distribution of ZnO nanoparticles and successful grafting onto the cellulose fibers (Figure 2e–i).

The structural evaluation on the prepared samples was obtained by XRD analyses. The diffractograms for cellulose, ZnO, and cellulose/ZnO composites are represented in Figure 3. ATcell and FTcell samples exhibited characteristic diffraction peaks of cellulose Iα (JCPDS Card No. 56-1719) with a shoulder containing overlapped signals at 14.7° and 16.2° (for (101) and (10$\bar{1}$) planes), a strong peak at 22.6° (002), and a weak signal at 34.3° (004) [18,26]. Cellulose extracted from almond trimmings (ATcell) and fig trimmings (FTcell) presented similar crystallinity values of 66.32% and 66.42%, respectively (Table 1). On the other hand, XRD patterns for the prepared ZnO sample displayed peaks at 31.68°, 34.32°, 36.15°, 47.41°, 56.46°, 62.71°, 67.79°, and 68.93° assigned to (100), (002), (101), (102), (110), (103), (112), and (201) characteristic planes of the hexagonal wurtzite structure of ZnO, respectively (JCPDS Card No. 36-1451). These peaks also presented in the diffractograms of ATcell/ZnO and FTcell/ZnO samples, proving that ZnO particles were successfully incorporated onto the cellulosic surface.

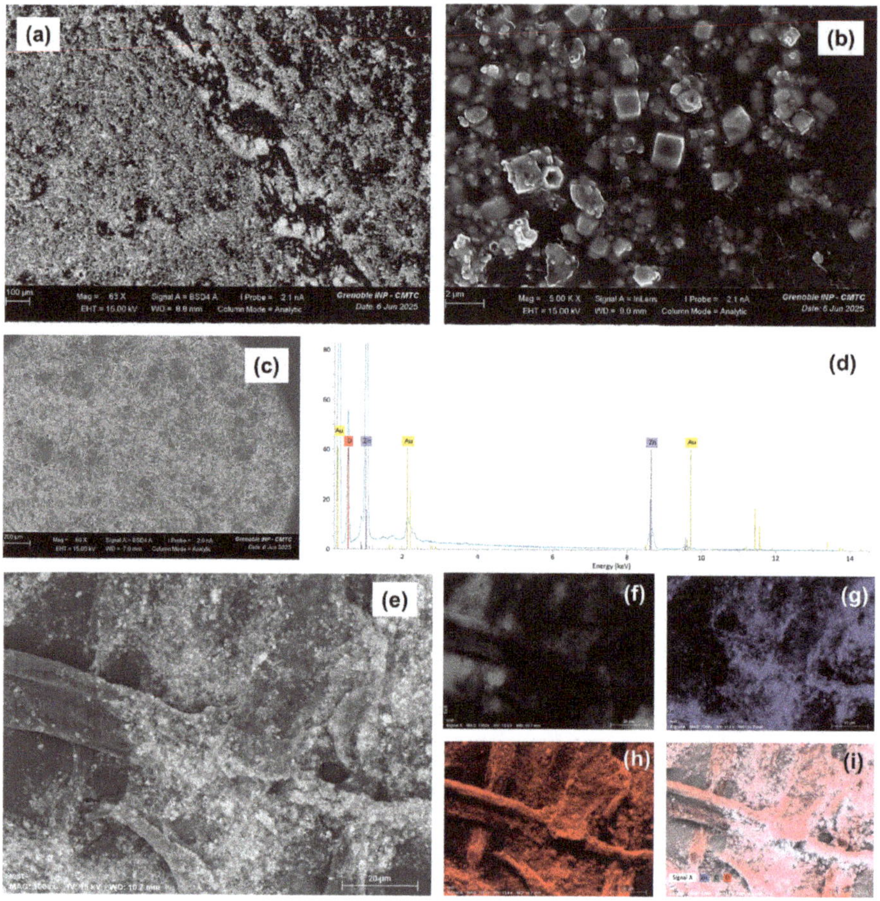

Figure 2. SEM imaging of ZnO (**a**,**b**) and FTcell/ZnO samples (**c**); elemental analysis spectrum of ZnO sample (**d**); EDX images (**e**); and mapping of ATcell/ZnO composite: C (**f**), Zn (**g**), O (**h**), and whole (**i**) (carbon in grey, zinc in blue and oxygen in red).

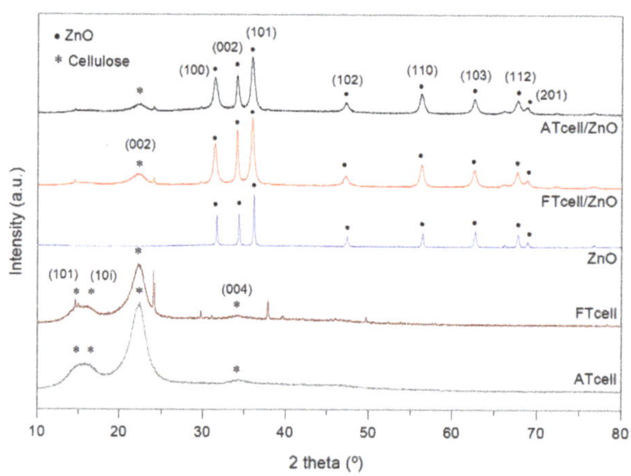

Figure 3. XRD patterns for ATcell, FTcell, ATcell/ZnO, ATcell/ZnO, and ZnO samples (peaks identified with * and with • for cellulose and ZnO, respectively).

Changes in the intensity and width of the diffraction peaks of ZnO grown on the cellulosic surface can be observed, which according to Zhang et al. [30] demonstrates that this substrate provides a large number of potential nucleation sites, reducing the stacking or agglomeration of ZnO particles. On the other hand, a significant decrease in cellulose-related peaks and crystallinity was observed for ATcell/ZnO and FTcell/ZnO composites (49.68% and 52.21%, respectively), being attributed to the presence of ZnO particles covering the cellulose surface and hampering the measurement [5,26]. According to Onyszko et al. [5], it was confirmed that the formation of zinc oxide was complete and without any traces of $Zn(OH)_2$ as intermediate product, since as they stated it presents some characteristic XRD peaks in the range of 10–30°, which were not found for our composites.

According to the Debye–Scherrer equation, the average size of ZnO crystallites in ZnO, ATcell/ZnO, and FTcell/ZnO samples was 43 ± 12 nm, 16 ± 3 nm, and 19 ± 7 nm, respectively (see Table S2). These results were in agreement with those of Li et al. [40], who reported similar particle sizes for ZnO powder (40.61 nm) and lower values for cellulose/ZnO composites (14–16 nm). In the literature, authors argued that the growth of ZnO on the cellulosic substrate allows for a better dispersion of oxide particles, which avoids agglomeration into a larger particle size and further a decrease in ZnO crystallites on the composite surface, improving photocatalytic performance [26,27,40]. In addition, the lattice strain ε was also calculated to evaluate possible distortion or deformation of the formed ZnO crystals. Thus, for the ZnO sample, an ε average value of $2.1 \pm 0.2 \times 10^{-3}$ was found (Table S2), whereas higher lattice strains of $5.6 \pm 1.6 \times 10^{-3}$ and $4.9 \pm 1.2 \times 10^{-3}$ were calculated for ATcell/ZnO and FTcell/ZnO samples, respectively. This strain increase is not considered severe and might be due to a reduction in particle size, a fact confirmed by the crystallite size values discussed above, but also to crystalline defects due to vacancies in the lattice or dislocations due to an interaction between the ZnO and the cellulose substrate that might produce these crystal deformations [28].

Adsorption/desorption isotherms (see Figure 4) and textural properties such as BET surface area and BJH average pore size (Table 1) of cellulose and cellulose/ZnO samples confirmed the success of ZnO particle formation onto cellulose surface. The original ATcell and FTcell samples were found to be macroporous materials, with low surface area of 1.8 and 3.3 m^2/g and few large pores (79.8 and 56.3 nm, respectively, with BJH cumulative pore volume of 0.05 cc/g). On the other hand, ATcell/ZnO, FTcell/ZnO, and ZnO samples presented a mesoporous behavior with 14.5, 14.9, and 13.4 m^2/g surface area and pores of 33.2, 36.1, and 38.1 nm in diameter, respectively.

Thus, the incorporation of ZnO allowed for an increase in the porosity of celluloses (up to 0.16 cc/g BJH cumulative pore volume), increasing the BET surface area, with more but smaller pores. However, the here-prepared ZnO material results were less porous than those obtained by Zafar et al. [25], even following a similar synthesis pathway, a condition that is considered key in the morphology and structure of ZnO particles, as it severely affects the nucleation and crystal growth, essential for controlling the content of the ZnO in the final composite and its photocatalytic behavior [18].

XPS survey spectra confirmed the presence of carbon (C 1s), oxygen (O 1s), and zinc in both ATcell/ZnO and FTcell/ZnO composites, as depicted in Figure 5a. For zinc, Zn 3d, Zn 3p, Zn 3s, several Zn LMN, and Zn 2p peaks were identified. Auger peaks occur when an excited atom releases energy without emitting a photon, which depends only on the electronic structure of the element and not on its chemical environment as in XPS peaks; thus, they help to distinguish between different oxidation states. In ZnO samples, an Auger peak located at 999 eV corresponds to samples containing ZnO (Zn^{2+}). In the spectrum, a peak close to 976 eV was also observed, which, although it could be attributed to a metallic Zn^0 state, is also characteristic of the KLL peak of oxygen in the O^{2-} state in metal oxide samples.

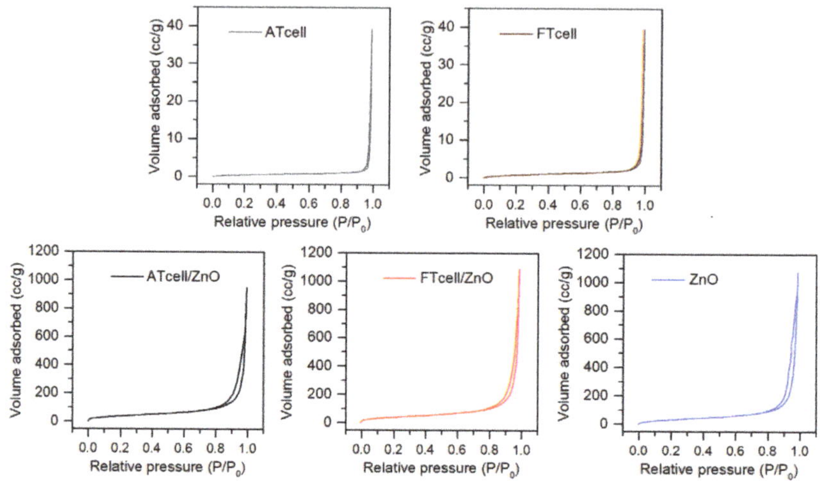

Figure 4. N_2 adsorption/desorption BET isotherms for the prepared samples.

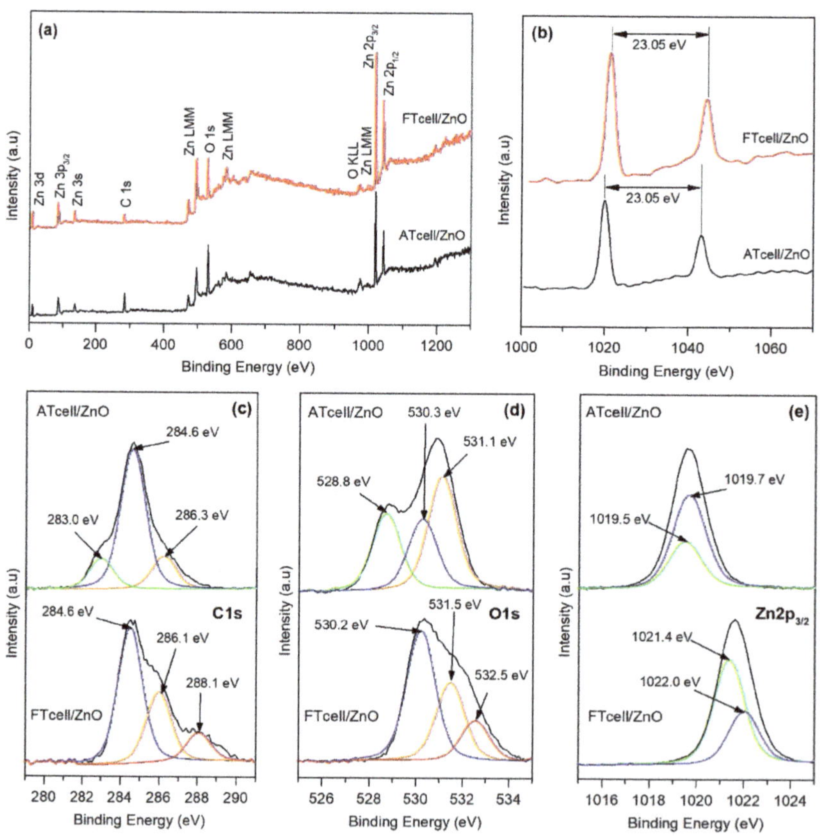

Figure 5. XPS survey scan full spectra of cellulose/ZnO composites (**a**), detailed spectra of Zn 2p signal (**b**), and high-resolution XPS spectra for (**c**) C 1s, (**d**) O 1s, and (**e**) Zn 2p3/2.

The signal doublet at ~1021 and ~1044 eV (Figure 5b), identified as the Zn 2p3/2 and Zn 2p1/2 lines, respectively, showed a characteristic spin-orbit separation of 23.05 eV, indicating a normal state of Zn^{2+} in ZnO [17,18]. The shift towards higher binding energies of the Zn 2p peaks has been observed before in the literature, being related to higher oxidation states of zinc [33] in the form of $Zn(OH)_2$ (which would indicate an incomplete formation of ZnO during the preparation of the composites) but also to a strong interaction of the ZnO particles with the hydroxyl groups of cellulose [27,43]. Nevertheless, in the present work, a shift towards lower binding energies was observed for the ATcell/ZnO (1019.8 and 1043.3 eV), which could indicate the coexistence of Zn^0 on the surface of the sample. The formation of metallic zinc is unlikely since no work has been reported where this occurs on any substrate without a calcination step during the preparation of nanoparticles [44], although the possible occurrence of changes in the normal error margin during the analysis has also been discussed [43].

The high-resolution XPS of C 1s for cellulose/ZnO composites demonstrated that several chemical states of C were present (Figure 5c), yielding characteristic binding-energy peaks at 284.6 eV and at around 286 eV assigned to C–C/C–H and C–O in the alcoholic/ether groups in the cellulose structure, respectively [26]. The lower binding 283.0 eV peak that appears for the ATcell/ZnO sample has been reported as the presence of carbon atoms in carbide form, indicating the formation of Zn–C bonds [44]. Thus, the formation of C–Zn interactions becomes more plausible than the possible formation of metallic zinc (lower energy shift observed above for the Zn 2p peak) on the cellulose surface. Cho et al. [44] corroborated the replacement of oxygen by carbon in the ZnO crystal structure by observing the lattice expansion in their XRD results. In the present work, higher values for the lattice strain than those previously reported for pure ZnO were found for both cellulose/ZnO composites (Table S2), being especially higher for ATcell/ZnO in the 002 plane, corresponding to the c-axis growth perpendicular to the substrate. On the other hand, for the FTcell/ZnO composite, a higher binding energy peak was recorded at 288.1 eV that could indicate the existence of C=O bonds in the surface of the sample, probably related to carboxyl groups [26,30], which agrees with the higher surface charge obtained above for the pristine FTcell sample.

The XPS spectrum of O 1s (Figure 5d) shows a lower binding peak at 528.8 eV for the ATcell/ZnO sample that indicates a strong interaction between oxygen and zinc forming the metal oxide. It also displayed a peak near 530 eV for both samples that has been reported as the characteristic O^{2-} ion in the hexagonal wurtzite structure of ZnO [29,31], thus confirming the formation of ZnO on both cellulosic samples. The peaks at around 531.1 and 531.5 eV for ATcell/ZnO and FTcell/ZnO, respectively, could be attributed to defects on the ZnO structure due to chemisorbed oxygen [27] and oxygen vacancies but also to the presence of C–O–Zn bonds [30], which corroborates the anchoring and growth of the ZnO particles in the hydroxyl groups of the cellulose surface [17]. The highest binding energy in the FTcell/ZnO spectrum (532.5 eV) is in fact due to the free C–OH groups on the substrate surface [24], indicating a more efficient nucleation of ZnO in the ATcell/ZnO sample, which presented lower ρ values and therefore a lower amount of hydroxyl groups on the surface. The measured atom % moieties (at ~10 nm dept on sample surface) by XPS analysis for carbon, oxygen, and zinc at their main peaks (located at around 286 eV for C 1s, at 530 eV for O 1s and at 1021 eV for Zn $2p_{3/2}$, respectively) resulted in 16.0% C, 22.5% O, and 61.5% Zn for ATcell/ZnO and 26.4% C, 26.4% O, and 47.2% Zn for FTcell/ZnO. All this confirmed that there was a clear difference in the composition and functionality between the original ATcell and FTcell samples, resulting in significant differences in the surface of the zinc oxide composites.

To evaluate the photocatalytic activity of a material, it is essential to know the range of the electromagnetic spectrum it can absorb and the minimum energy required to excite electrons from the valence band to the conduction band, i.e., the bandgap. The optical response of the prepared materials is shown in Figure 6.

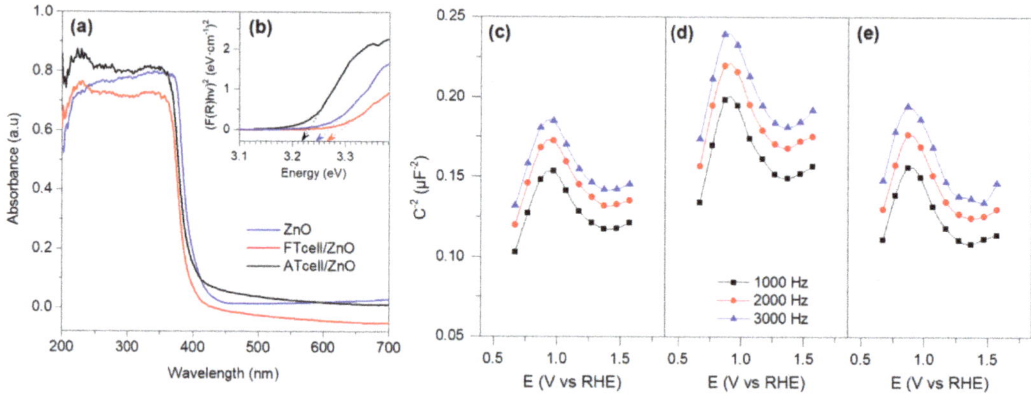

Figure 6. UV–vis DRS spectra of ZnO and cellulose/ZnO composites (**a**), Tauc plots of bandgap energy calculation (**b**), and Mott–Schottky plots of ZnO (**c**), ATcell/ZnO (**d**), and FTcell/ZnO (**e**) samples.

By comparing the prepared ZnO particles and cellulose/ZnO composites, the light absorption intensity of the cellulose/ZnO composites increased, and it was found that the cellulose/ZnO composites revealed excellent UV absorption properties in the visible region. As shown in Figure 6a, the absorption edge of the cellulose/ZnO composites was found to shift below 400 nm, indicating a decrease in the bandgap energy of the pure ZnO.

As illustrated in Figure 6b, bandgap energies (E_g, eV) of the ZnO, ATcell/ZnO, and FTcell/ZnO composites were estimated to be 3.24, 3.22, and 3.27 eV, respectively. Thus, all the analyzed materials presented a bandgap close to 3.2 eV, indicating that they are suitable to work in the UV light range so that the photogenerated electron–hole pair would react with OH groups (from water and/or the surface of the materials) to achieve the degradation of contaminants in an aqueous medium.

The bandgap value obtained for the prepared ZnO particles is in agreement with the results reported in the literature [21,23,30,44], although this energy depends greatly on the size of the metal oxide particles. Li et al. [26] also observed a small decrease in the bandgap energy for cellulosic composites with zinc oxide compared to pristine ZnO particles, which they attributed to the surface charge of the initial fiber that improves the deposition of zinc oxide. Likewise, as Cho et al. suggest in their work [44], certain interactions with doping materials in vacancies of the ZnO crystal structure would produce a reduction in the absorption energy and bandgap of the material. Thus, crystallites with vacancies and defects in their structure could enhance the interaction of ZnO with light radiation (hv), improving its photocatalytic activity since oxygen vacancies generate energy levels within the ZnO bandgap, facilitating the absorption of lower-energy photons. This can broaden the spectral response of ZnO particles into the visible region, improving their activity under sunlight. Here, the greater presence of O^{2-} vacancies in the crystalline structure of the ZnO particles, previously observed in the XPS results, together with a lower agglomeration of the same according to the XRD measurements carried out, could explain the reduction of the bandgap for the ATcell/ZnO sample. In this way, cellulose composites demonstrated good visible light harvesting capabilities compared to ZnO particles, probably due to the charged and OH-rich surface of cellulose acting as a sensitizer in bandgap modulation on composites.

According to the Mott–Schottky analysis and as can be seen in Figure 6c,d, all the materials used in this work present the typical behavior of an n-type semiconductor. In this way, we can approximate the conduction band limit of the different semiconductors. The flat band (where C-2 becomes 0 V) of ZnO is more negative (0.1 V vs. RHE) than those of the ATcell/ZnO and FTcell/ZnO samples (0.2 V vs. RHE), which is a shift to positive potentials of 0.1 V. This now allows us to approximate the conduction band of ZnO to −3.77 eV and that of the ZnO-modified celluloses to −3.87 eV [45]. Due to the fact that this difference in potential in the conductivity band of the different samples is very small, it cannot be related to a higher degradation of the different dyes studied in this work, which could mean that the higher degradation of the dyes is due to a smaller particle size in the ZnO-modified cellulose samples than the pristine ZnO itself.

3.2. Dye Removal Performance

As confirmed in the literature, the preparation of cellulosic composites with zinc oxide depends strongly on the methodology followed, which affects the morphology and structure of the ZnO particles [12] but also the existence of –OH groups available on the substrate surface that acts as a driving force during ZnO nucleation due to electrostatic interactions [28]. A mechanism of formation and action is described below (Equations (3)–(9) and Figure 7).

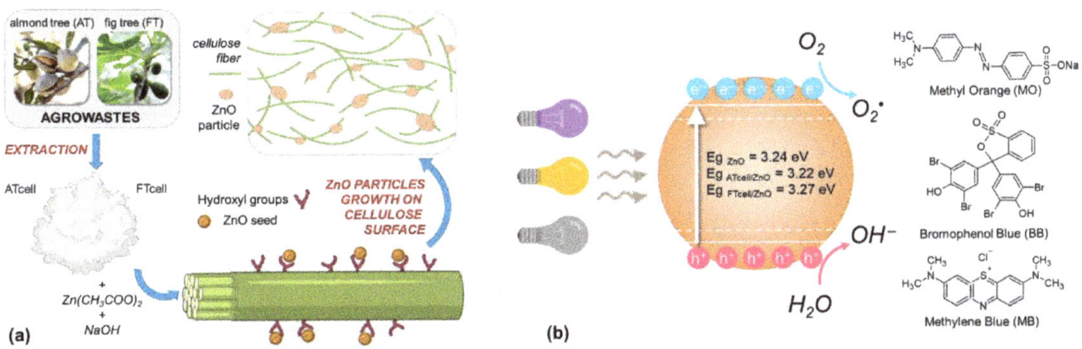

Figure 7. The proposed mechanisms for cellulose/ZnO composite synthesis (**a**) and for the photodegradation of the different studied dyes (**b**).

In general, as shown in Equation (3), the use of zinc acetate as a precursor for the growth of ZnO particles begins with the formation of Zn^{2+} ions, which interact strongly with the –OH present in the medium but also on the surface of the cellulose (Figure 7a). Thus, the formation of hydroxide $Zn(OH)_2$ is forced [25], and when the medium is saturated with –OH groups of alkaline origin, the formation of $Zn(OH)_4{}^{2-}$ (Equation (4)) occurs on the surface and interstice of the fibers [26], which subsequently decomposes (Equation (5)), initiating the nucleation of ZnO [5].

$$Zn^{2+} + 2\,OH^- \rightarrow Zn(OH)_2 \tag{3}$$

$$Zn(OH)_2 + 2\,OH^- \rightarrow Zn(OH)_4^{2-} \tag{4}$$

$$Zn(OH)_4^{2-} \rightarrow ZnO + H_2O + 2\,OH^- \tag{5}$$

Subsequently, the catalytic activity of ZnO is based on the creation of electron–hole pairs from the interaction with light (Equation (6)), achieving the generation of active species $OH^•$ and $O_2{}^•$ in the aqueous media (Equations (7) and (8)) that interact with the pollutants [13,21] and degrade them (Figure 7b) into non-toxic and simpler compounds (Equation (9)). Thus, as previously discussed, oxygen vacancies could increase the affinity of ZnO for molecules such as O_2 and H_2O, promoting the formation of reactive species [20], such as $OH^•$ and $O_2{}^•$, which are key in photocatalytic processes. Furthermore, it could be assumed that structural defects in ZnO would improve electron mobility in the crystal lattice, reducing recombination and increasing the efficiency of free radical generation.

$$\mathrm{ZnO} + \mathrm{h}\nu \rightarrow \mathrm{ZnO}\left(\mathrm{h}^+ + \mathrm{e}^-\right) \tag{6}$$

$$\mathrm{OH}^- + \mathrm{h}^+ \rightarrow \mathrm{OH}^\bullet \tag{7}$$

$$\mathrm{O}_2 + \mathrm{e}^- \rightarrow \mathrm{O}_2{}^\bullet \tag{8}$$

$$\mathrm{dye} + \mathrm{OH}^\bullet + \mathrm{O}_2{}^\bullet \rightarrow \mathrm{degradation\ products} \tag{9}$$

To obtain the best comparative study, the photocatalytic activities of ZnO and cellulose/ZnO composites against various dyes were evaluated. It should be noted that although the dosage during these studies was 0.1 g of composite in each experiment, for ATcell/ZnO and FTcell/ZnO, approximately only 47% wt. of the sample corresponded to ZnO particles (as seen in Table 1). For this reason and in order to better evaluate the possible dual adsorptive/photocatalytic mechanism of the samples, the specific removal per ZnO active site (Rs, % removal/g ZnO) was defined as follows:

$$\mathrm{Rs} = \frac{\mathrm{R}}{\mathrm{IM}} \times 100 \tag{10}$$

where R (%) is the dye removal achieved by tested mass g of the sample, calculated according to Equation (2), and IM (%) is the inorganic matter for the prepared material as calculated by Equation (1) and appeared in Table 1.

As shown in Figure 8, there was a remarkable difference in dye removal efficiency across different light sources and composites. Notably, in most cases, the maximum dye degradation occurred within the first 5 min of exposure with the samples, without observing a progressive removal enhancement over time, which could suggest the saturation or inactivation of the catalyst surface by the dye molecule [30].

Figure 8a shows significant photocatalytic activity of ZnO on the removal of the MO dye under the conditions studied, as a maximum specific reduction of 51, 43, and 26% removal/g ZnO of the dye was achieved under UV light, white light, and no light irradiation. In the work of Rodwihok et al. [46], pristine ZnO particles (95 nm of crystal size) barely managed to adsorb MO in the dark (<5% removal) and reached approximately 20% of dye elimination after 20 min of sunlight irradiation. Zafar et al. [25] reported the adsorptive capacity of ZnO nanospheres prepared by the analogous alkaline precipitation method but with better textural properties (S_{BET} 49.36 m^2 g^{-1} and D_{BJH} 27.44 nm) and where a dose of 0.1 g of nanoparticles managed to adsorb about 50% of MO under similar experimental conditions (pH 6 and room temperature). Thus, the importance of particle size on the adsorptive/photocatalytic removal capacity of ZnO resulted evident. The ATcell/ZnO and FTcell/ZnO cellulosic composites also showed certain yield sensitivity to light exposure, improving the specific MO removal capacity observed for pure ZnO by about 30% under all three experimental conditions. As discussed in the sample characterization, both cellulose/ZnO composites exhibited better textural (8–11% higher S_{BET} surface area and 5–13% lower average D_{BJH} pore size) and structural (56–63% reduction in D crystallite size) properties than the as-prepared ZnO particles. Thus, although the amount of zinc oxide present is lower in ATcell/ZnO and FTcell/ZnO, the improved surface area provided by the cellulose substrate for uniform and distributed particle growth had a tremendous positive impact on the MO dye removal capacity. By comparing the results of experiments carried out under UV light and those carried out in the dark, it can be confirmed that the removal of MO with the materials studied is effective following an adsorption mechanism and that it is enhanced with the application of white or ultraviolet light.

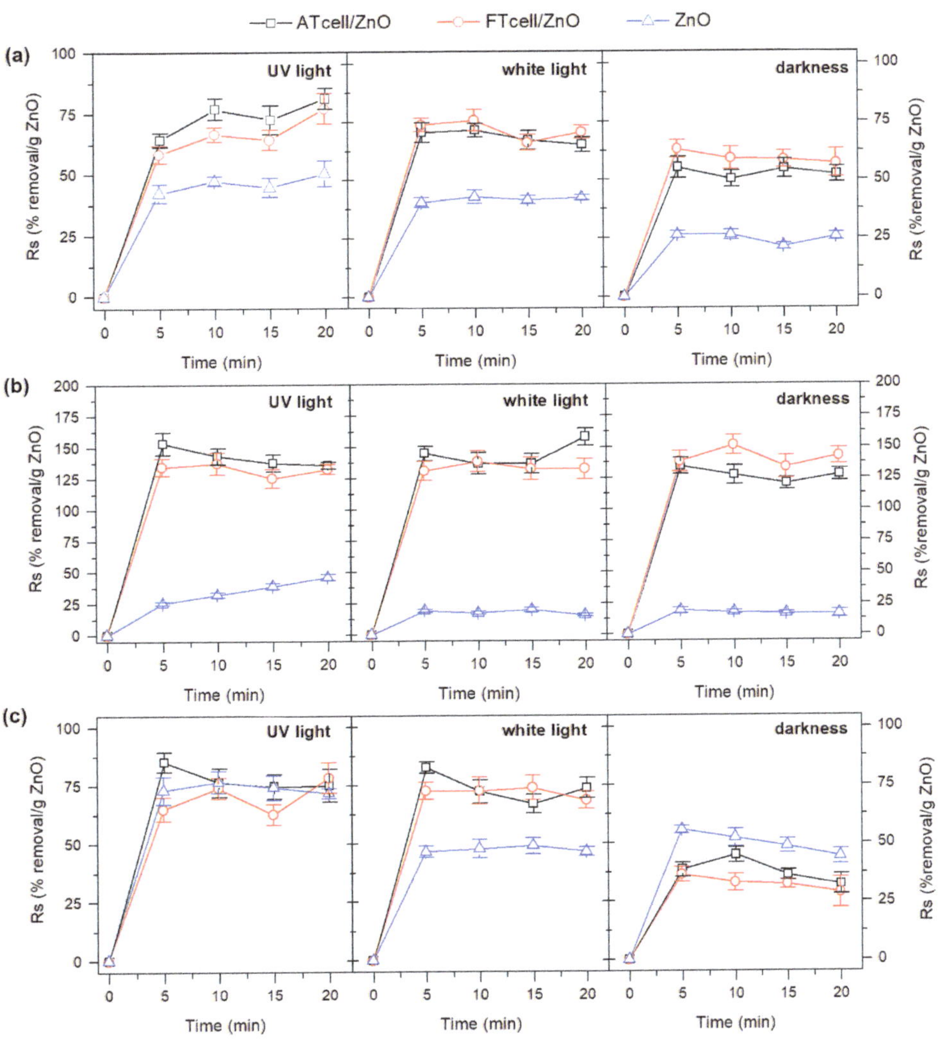

Figure 8. The performance of the ATcell/ZnO, FTcell/ZnO, and ZnO samples on the specific removal (Rs, % removal/ZnO content) of (**a**) MO, (**b**) MB, and (**c**) BB dyes.

Similarly, for MB (cationic dye) removal assays, the adsorption/desorption behavior of the materials becomes a key factor, demonstrating the great influence of the catalysts surface on the MB dye removal efficiency. This is confirmed by the results shown in Figure 8b with a predominant photocatalytic effect of ZnO on the removal of MB, with a maximum around 15–17% of dye removal/ZnO content under white light or in the dark, while irradiation with UV light achieved close to 25% removal after 5 min of the experiment, increasing linearly up to 45% removal after a 20 min assay. The ATcell/ZnO and FTcell/ZnO composites showed good behavior in experiments conducted under three light conditions, removing near to 73% of the dye after 5 min of assay (with Rs up to 150%). As indicated by Dehghani et al. [37], the degradation of MB on cellulosic composites with ZnO requires a first adsorption step, allowing the contaminants to interact with the catalyst to facilitate the photodegradation process. These authors reported a maximum of 20% MB degradation after 20 min of solar or UV radiation using ZnO particles (12.15 m^2/g surface area, 23 nm pore size), while when supported on cellulose nanofibers (with a relative ZnO loading of 29–72% wt.), they achieved 30–95% degradation. For ZnO concentrations similar to those found in ATcell/ZnO and FTcell/ZnO (around 47% wt according to Table 1), the aforementioned work [37] achieved less than 40% MB removal in 20 min (using IZnOCNF sample with 2.24 m^2/g surface area and 63 nm pore diameter) but over 80% degradation with a composite with better textural properties (using VIZnOCNF with 10.94 m^2/g and 7.9 nm pore diameter). In fact, adsorption proved to be the main mechanism for this dye removal by the ATcell/ZnO and FTcell/ZnO samples, as seen in Figure 8b (experiments irradiated with light or in the dark).

Again, the dual adsorptive/photocatalytic effect of ZnO was also clear, as shown in Figure 8c, where a better adsorption of BB on the ZnO surface could be observed (between 44 and 47% Rs under white light or in darkness after 20 min), promoted by the possible ionization of Zn-OH groups on the catalyst surface, which also achieves a degradation of about 72% Rs of BB after 20 min when the experiment is irradiated with UV light. There are few studies reporting results on the potential adsorption and photodegradation of BB from aqueous effluents [29]. The rapid and complete removal of BB (12 min under visible light) has been reported [47] using a combined system of a recoverable biogenic hematite photocatalyst (G-Fe_2O_3-NP's) with the oxidizing agent H_2O_2. Another example is the work of Akpomie et al. [48], where only the adsorptive capacity of ZnO nanoparticles was evaluated under different conditions (pH, dye concentration, ZnO dosage, temperature, and previous sonication). Thus, in similar experiments, these authors found a 50–70% removal of BB after 180 min of testing using loads of 0.1–0.3 g of ZnO nanoparticles, dropping to approximately 20% removal in 20 min. In the present work, the ATcell/ZnO and FTcell/ZnO composites achieved a net adsorption (dark tests) of 13–15% of the dye (Rs of 28–32% removal/g ZnO in Figure 8c), improving to up to 32–37% removal after 20 min under UV and white light irradiation (Rs of 69–79% BB removal/ZnO content in the sample). Lower yields of 20% in BB photodegradation have been reported by Shah et al. [49] after 2 h of UV light exposure using graphene nanoplate-supported titanium oxide (TiO_2/GNP). However, no references have been found to studies on the adsorption/photodegradation processes for BB using cellulose composites with zinc oxide, the present work being the first to report results.

The reusability study was performed for the BB removal process irradiated with UV light. As shown in Figure 9, both composites showed adequate behavior in the second dye removal cycle, with the Rs value slightly reducing from 75.11 ± 6.99% to 73.28 ± 3.92% for the ATcell/ZnO composite and from 78.47 ± 6.55% to 77.45 ± 2.96% when FTcell/ZnO was used as the photo-adsorbent agent.

After the third reuse, the efficiency decreased slightly for both samples, resulting in Rs values of 60.83–4.30% and 66.70–2.76% for ATcell/ZnO and FTcell/ZnO, respectively; that is, after three uses, the composites lost 19% and 15% of their initial capacity to remove BB dye under UV light. Other authors did not observe loss of stability or activity of their cellulosic composites with ZnO [26], while others found only a 10% loss of efficacy after 5 reuse cycles [13]. The reusability of this type of hybrid material with photo-adsorptive capacity is very important for its application in wastewater treatment processes, and it is considered necessary to study various eluents that can regenerate the composites without affecting their stability and effectiveness [29,39].

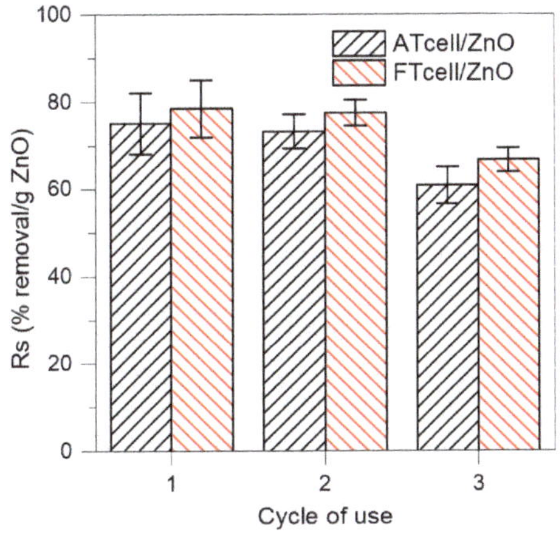

Figure 9. Specific removal (Rs, % removal/ZnO content) reached by regenerated and reused ATcell/ZnO and FTcell/ZnO composites for BB removal experiments (20 min contact under UV light irradiation).

3.3. Understanding the Effect of Cellulose/ZnO Composite Properties in Dye Removal Success

The multivariate statistical analysis performed on the set of results obtained (Tables S3–S5 in SI) allowed us to distinguish significant differences in the behavior of the materials during the removal of the three dyes considered. In this way, it was possible to verify that the structural and textural properties evaluated for the tested materials showed a significant relationship with each other, as shown in Figure 10.

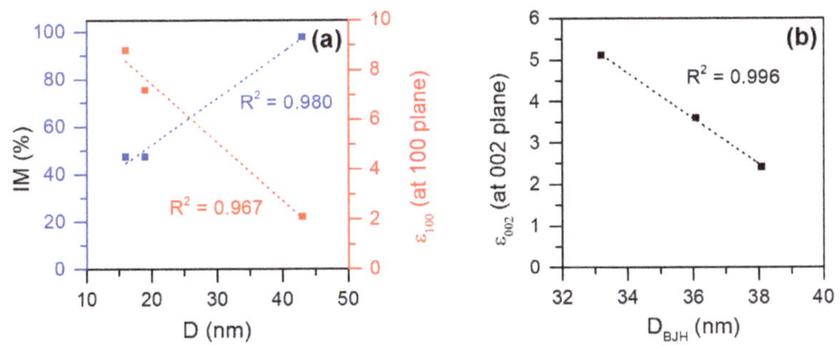

Figure 10. Linear fit analyses for parameter pairs. Relationship between compositional and structural characteristics on the crystallite size D of the grown ZnO particles (**a**) and on the average pore diameter D_{BJH} of the prepared samples (**b**).

Figure 10a shows the most significative relationships found, with ZnO crystallite size (D) strongly and directly related with the inorganic load (IM) present in ZnO, ATcell/ZnO, and FTcell/ZnO samples (see Table 1), avoiding ZnO particle agglomeration when cellulose was used as support. An indirect relationship of crystallite size was also observed with the lattice strain at the 100 plane (Table S2 in SI). As discussed in the literature [28], the 100 plane of ZnO has a higher surface energy than others, and it is in this plane where there is greater interaction between ZnO and cellulose [33], avoiding particle growth in the direction of the c plane and limiting their size. Thus, the higher the ε_{100}, the smaller the diameter D of the ZnO crystals formed. These authors also indicated that the electrostatic interactions between ZnO and cellulose, which cause a slower growth of the particles, force the development along the surface of the cellulose, affecting the 002 plane of the zinc oxide, thus favoring a uniform superficial growth, affecting the textural properties of the material. This can be seen in Figure 10b, where the strong relationship between the lattice strain ε_{002} and the average pore diameter D_{BJH} is observed, such that the higher tensions in the formed crystals translate into a less compacted and dispersed growth, improving the porosity of the surface.

The combined multivariable analysis of all experiments showed no significant effect of evaluated properties on elimination extent, so an individual study was conducted for each dye. Interestingly, surface charge ρ was strongly associated with the extent of pollutant removal from the experiments performed, and more specifically with the type of dye, as shown in the fitted curve graphs in Figure 11. It can be seen how the presence of positive charges, due to the protonation of the hydroxyl groups present on the surface of ZnO, strongly influences any type of applied light, which clearly makes the surface charge a key factor in the removal mechanisms. Furthermore, while for MO and BB, the R-ρ relationship is direct, for MB, it is strongly indirect, which is confirmed by the cationic nature of this dye and the stronger repulsive interactions with the surface of the ZnO composite [29]. Under UV light irradiation (Figure 11a), the removal efficiency was greater for BB (red dot curve), a weakly charged molecule in its anionic form that would have a sulfonate group and an alkoxide in its structure (Table S1), while MO (black square curve) only has one sulfonate group, so its possible interaction with the surface of the composite is reduced (less steep slope in the fitted curve). In addition, BB is known to absorb in the UV light range, making the removal observed for this dye more pronounced due to the photocatalytic nature of the process.

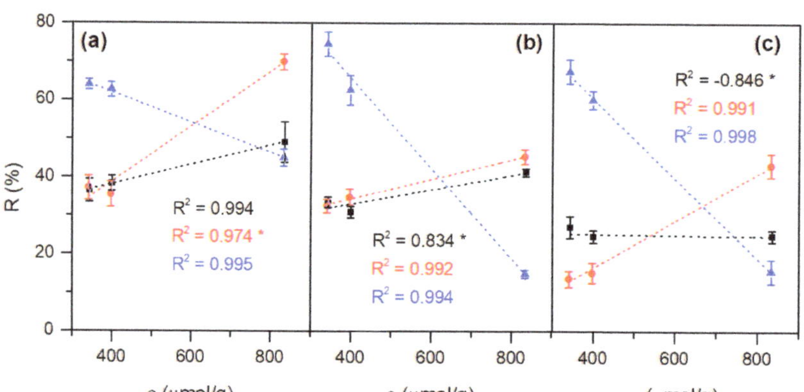

Figure 11. Fitted curves of removal extent (R, %) and surface charge of the prepared materials (ρ, μmol/g) after 20 min exposure with MO (black square), BB (red dot), or MB (blue triangle) under UV light (**a**), white light (**b**), or in darkness (**c**). For each curve, the adjusted R^2 is presented and * denotes that the slope is NOT significantly dependent on data at 95% confidence ($p > 0.05$).

When experiments were carried out under white light (Figure 11b), MO experienced almost the same degradation as under UV radiation, and the removal performance for BB was comparable to the MO one, both dyes being of anionic character. The MB retained a strong indirect relationship between removal and the surface charge of the composite used, but under white light, this relationship was even stronger (steeper curve slope), suggesting that the adsorption of the dye on the surface of the ZnO-prepared materials was better.

Likewise, MB removal in darkness (blue triangle in Figure 11c) followed the same behavior as under visible light, obeying mainly an adsorption mechanism highly influenced by the surface charge of the material, being much better for prepared cellulose/ZnO composites than for pristine ZnO particles. While for MO almost no influence of the surface charge of the material on the removal success was distinguished, the BB dye experienced the lowest removal values in darkness, although directly dependent on the positive charge of the adsorbent surface, which clarifies the potential combined use of photocatalysts/adsorbents in processes for the elimination of this pollutant, which has not been studied to date and opens an interesting research scope for the application of hybrid cellulosic materials with photocatalytic properties.

4. Conclusions

In this study, ZnO nanoparticles were first synthesized onto the cellulose fibers surface using a chemical in situ method. The prepared samples were subsequently characterized, demonstrating a significant improvement in cellulose surface texture thanks to the presence of ZnO particles (up to eight-fold higher S_{BET} and up to 60% reduction in D_{BJH} than the initial cellulose) as well as the appearance of optical properties similar to those of ZnO (bandgap close to 3.2 eV and performance in the UV and visible range). In this way, it was possible to confirm that the prepared composites could have potential applications as adsorbents/photocatalysts. Both cellulose/ZnO composites demonstrated good performance for specific removal of MO, MB, or BB pollutants, driven by the synergistic effects of adsorptive and photocatalytic mechanisms. Thus, the ATcell/ZnO and FTcell/ZnO composites showed a preferential MB removal mechanism by adsorption (up to 75% in 20 min), mainly caused by electrostatic attraction between the dye and the adsorbent surface. For the anionic dyes MO and BB, the dual adsorption/photodegradation performance of the prepared composites had evident results, with different behaviors observed depending on the type of light applied. This demonstrated how the use of a suitable support, such as cellulose, can significantly improve the dispersion of ZnO particles (smaller particles and prevent agglomeration), with clear improvements in their photocatalytic activity and surface properties. The obtained results for MO removal were similar to some found in the literature (with approximately 40% removal after 20 min of UV light). However, no references have been found to studies on the adsorption/photodegradation processes for BB using cellulose composites with zinc oxide, the present work being the first to report some results (about 35% elimination after 20 min of white or UV light). This study highlights the potential of cellulose/ZnO composites as highly effective photocatalysts for wastewater treatment in the textile industry. Future research could focus on the comparison of the composites' formulation, reusability, and testing effectiveness against other dye and organic pollutants to further advance environmental remediation.

Supplementary Materials: The following supporting information can be downloaded at: https://www.mdpi.com/article/10.3390/polym17131737/s1, Table S1: Dyes used for the evaluation of adsorptive/photocatalytic efficiency of cellulose/ZnO composites; Text S1: Characterization methods; Text S2. Removal of dyes under controlled light irradiation; Figure S1: Scheme of the photoreactor prototype used in this work (a) and recorded spectra for the employed UV (b) and white (c) light sources; Figure S2: Optical Microscopy observations of ZnO particles and cell/ZnO composites; Table S2: X-ray diffraction planes considered during crystallite size (D, nm) and lattice strain (ε) determination for ZnO particles present in the prepared samples; Text S3. Statistical analysis; Table S3: Statistically determined correlation parameters for variables considered during methyl orange (MO) removal; Table S4: Statistically determined correlation parameters for variables considered during methylene blue (MB) removal; Table S5: Statistically determined correlation parameters for variables considered during bromophenol blue (BB) removal. References [35,36] are cited in the Supplementary materials.

Author Contributions: J.B.: Investigation, Writing—original draft. R.K.: Methodology, Writing—review and editing, Validation, Supervision. V.G.-C.: Investigation, Writing—review and editing. A.A.R.: Formal analysis, Validation, Funding acquisition. A.G.: Conceptualization, Methodology, Writing—review and editing, Supervision, Validation, Funding acquisition. All authors have read and agreed to the published version of the manuscript.

Funding: Financial support was provided by the University of Cordoba through the "Plan Propio de Investigación Enrique Aguilar Benítez de Lugo 2023, Submodalidad 2.4 UCOLIDERA" awarded to AgroCell project and by the PCI2024-153418 project, funded by MICIU/AEI/10.13039/501100011033 and co-funded by the European Union under the PRIMA 2023 call. The APC was funded by PAIDI group FQM-383 of the University of Cordoba.

Data Availability Statement: The original contributions presented in this study are included in the article/Supplementary Material. Further inquiries can be directed to the corresponding author.

Acknowledgments: Authors gratefully acknowledge the financial support provided by the University of Monastir, Tunisia, through the scholarship of Jihene Belhaj, by the University of Cordoba for the financial support provided through the "Plan Propio de Investigación Enrique Aguilar Benítez de Lugo 2023, Submodalidad 2.4 UCOLIDERA" awarded to AgroCell project, and by the DurInnPack PCI2024-153418 project, funded by MICIU/AEI/10.13039/501100011033 and co-funded by the European Union under the PRIMA 2023 call. Valentín García-Caballero thanks the "Plan Propio de Investigación" from the University of Córdoba for financial support through a predoctoral contract (Submodality 2.2). The authors express their sincere gratitude to Jean-Luc Putaux (CERMAV, Grenoble) and Karine Janel (LGP2, France) for their valuable support in conducting the morphological analyses and to the research group FQM-346 "Organic Catalysis And Nanostructured Materials" of the Organic Chemistry Department of the University of Cordoba, especially to Dolores G. Gil-Gavilán for her help during the UV-vis DRS analysis performed.

Conflicts of Interest: The authors declare that they have no known competing financial interests or personal relationships that could have influenced the results or interpretations presented in this work.

Abbreviations

The following abbreviations are used in this manuscript:

AT	Almond tree
ATcell	Cellulose extracted from almond tree
BB	Bromophenol blue
CrI	Crystalline index (%)
D	Crystallite size (nm)
D_{BJH}	average pore diameter ()
Eg	Bandgap (eV)
FT	Fig tree
FTcell	Cellulose extracted from fig tree
IM	Inorganic matter (% wt.)
L	Fiber length (mm)
MB	Methylene blue
MO	Methyl orange
R	Dye removal (%)
Rs	Specific dye removal (% removal/g ZnO)
S_{BET}	Surface area (m^2/g)
W	Fiber width (μm)
ε_{002}	Lattice strain at 002 plane in ZnO structure
ε_{100}	Lattice strain at 100 plane in ZnO structure
ρ	Surface charge (μmol/g)

References

1. Bailey, K.; Basu, A.; Sharma, S. The Environmental Impacts of Fast Fashion on Water Quality: A Systematic Review. *Water* **2022**, *14*, 1073. [CrossRef]
2. Khadhri, N.; El Khames Saad, M.; Ben Mosbah, M.; Moussaoui, Y. Batch and Continuous Column Adsorption of Indigo Carmine onto Activated Carbon Derived from Date Palm Petiole. *J. Environ. Chem. Eng.* **2019**, *7*, 102775. [CrossRef]
3. Crini, G. Non-Conventional Low-Cost Adsorbents for Dye Removal: A Review. *Bioresour. Technol.* **2006**, *97*, 1061–1085. [CrossRef]
4. Tan, S.N.; Yuen, M.L.; Ramli, R.A. Photocatalysis of Dyes: Operational Parameters, Mechanisms, and Degradation Pathway. *Green Anal. Chem.* **2025**, *12*, 100230. [CrossRef]
5. Onyszko, M.; Markowska-Szczupak, A.; Rakoczy, R.; Paszkiewicz, O.; Janusz, J.; Gorgon-Kuza, A.; Wenelska, K.; Mijowska, E. The Cellulose Fibers Functionalized with Star-like Zinc Oxide Nanoparticles with Boosted Antibacterial Performance for Hygienic Products. *Sci. Rep.* **2022**, *12*, 1321. [CrossRef] [PubMed]
6. Dhawan, A.; Sudhaik, A.; Raizada, P.; Thakur, S.; Ahamad, T.; Thakur, P.; Singh, P.; Hussain, C.M. $BiFeO_3$-Based Z Scheme Photocatalytic Systems: Advances, Mechanism, and Applications. *J. Ind. Eng. Chem.* **2023**, *117*, 1–20. [CrossRef]
7. Kumar, R.; Sudhaik, A.; Sonu; Nguyen, V.-H.; Van Le, Q.; Ahamad, T.; Thakur, S.; Kumar, N.; Hussain, C.M.; Singh, P.; et al. Graphene Oxide Modified K, P Co-Doped g-C_3N_4 and $CoFe_2O_4$ Composite for Photocatalytic Degradation of Antibiotics. *J. Taiwan Inst. Chem. Eng.* **2023**, *150*, 105077. [CrossRef]
8. Maruthupandy, M.; Rajivgandhi, G.; Muneeswaran, T.; Vasantharaj, S.; Sandoval-Hevia, G.; Muneeswaran, M.; Radzi, N.A.B.M. Cross-Linked Graphene with Chitosan Nanocomposites for Efficient Photocatalytic Degradation of Bromothymol Blue, Bromophenol Blue Dye Molecules. *Int. J. Biol. Macromol.* **2025**, *307*, 142132. [CrossRef]
9. Aldaz, B.; Figueroa, F.; Bravo, I. Cellulose for the Effective Decontamination of Water Pollution. *Rev. Bionatura* **2020**, *5*, 1150–1155. [CrossRef]
10. Fukuhara, M.; Yokotsuka, T.; Takashina, T.; Fujima, N.; Morita, M.; Ito, T.; Nakatani, T.; Hashida, T. High-Energy Storage Capacity of Cellulose Nanofiber Supercapacitors Using Bound Water. *Sci. Rep.* **2023**, *13*, 16600. [CrossRef]
11. Oyewo, O.A.; Elemike, E.E.; Onwudiwe, D.C.; Onyango, M.S. Metal Oxide-Cellulose Nanocomposites for the Removal of Toxic Metals and Dyes from Wastewater. *Int. J. Biol. Macromol.* **2020**, *164*, 2477–2496. [CrossRef] [PubMed]
12. Shi, C.; An, B.; Zhang, L.; Zai, Z.; Shi, Z.; Wang, Z.; Ma, J. Contribution of Surface Carboxyl of Cellulose in the Formation Mechanism and Interfacial Catalysis Activity of ZnO/Cellulose Nanocomposites. *Appl. Surf. Sci.* **2023**, *618*, 156633. [CrossRef]
13. Li, Y.; Yan, M.; Li, X.; Ma, J. Construction of Cu_2O-ZnO/Cellulose Composites for Enhancing the Photocatalytic Performance. *Catalysts* **2024**, *14*, 476. [CrossRef]
14. da Silva-Neto, M.L.; de Oliveira, M.C.A.; Dominguez, C.T.; Lins, R.E.M.; Rakov, N.; de Araújo, C.B.; Menezes, L.d.S.; de Oliveira, H.P.; Gomes, A.S.L. UV Random Laser Emission from Flexible ZnO-Ag-Enriched Electrospun Cellulose Acetate Fiber Matrix. *Sci. Rep.* **2019**, *9*, 11765. [CrossRef] [PubMed]
15. Musino, D.; Rivard, C.; Landrot, G.; Novales, B.; Rabilloud, T.; Capron, I. Hydroxyl Groups on Cellulose Nanocrystal Surfaces Form Nucleation Points for Silver Nanoparticles of Varying Shapes and Sizes. *J. Colloid Interface Sci.* **2021**, *584*, 360–371. [CrossRef]
16. Somogyi Škoc, M.; Macan, J.; Jakovljević, S.; Rezić, I. Synthesis and Characterization of ZnO and TiO_2 Hybrid Coatings for Textile UV Anti-Aging Protection. *Polymers* **2024**, *16*, 2001. [CrossRef]
17. Fu, F.; Li, L.; Liu, L.; Cai, J.; Zhang, Y.; Zhou, J.; Zhang, L. Construction of Cellulose Based ZnO Nanocomposite Films with Antibacterial Properties through One-Step Coagulation. *ACS Appl. Mater. Interfaces* **2015**, *7*, 2597–2606. [CrossRef]
18. Ma, J.; Zhu, W.; Min, D.; Wang, Z.; Zhou, X. Preparation of Antibacterial Self-Reinforced Zinc Oxide-Cellulose Composite by the Synthesis of ZnO in Partially Dissolved Cellulose. *Cellulose* **2016**, *23*, 3199–3208. [CrossRef]
19. Elfeky, A.S.; Salem, S.S.; Elzaref, A.S.; Owda, M.E.; Eladawy, H.A.; Saeed, A.M.; Awad, M.A.; Abou-Zeid, R.E.; Fouda, A. Multifunctional Cellulose Nanocrystal /Metal Oxide Hybrid, Photo-Degradation, Antibacterial and Larvicidal Activities. *Carbohydr. Polym.* **2020**, *230*, 115711. [CrossRef]
20. Liu, M.; Ye, Y.; Xu, L.; Gao, T.; Zhong, A.; Song, Z. Recent Advances in Nanoscale Zero-Valent Iron (nZVI)-Based Advanced Oxidation Processes (AOPs): Applications, Mechanisms, and Future Prospects. *Nanomaterials* **2023**, *13*, 2830. [CrossRef]
21. Luu, T.V.H.; Nguyen, H.Y.X.; Nguyen, Q.T.; Nguyen, Q.B.; Nguyen, T.H.C.; Pham, N.C.; Nguyen, X.D.; Nguyen, T.K.; Dao, N.N. Enhanced Photocatalytic Performance of ZnO under Visible Light by Co-Doping of Ta and C Using Hydrothermal Method. *RSC Adv.* **2024**, *14*, 12954–12965. [CrossRef] [PubMed]
22. Xiao, L.; Ma, J.; Zou, H.; Song, X.; Zhu, Q.; Chen, J.; Tang, X.; Zhou, L.; Chen, J.; Liu, Z.; et al. Multifunctional Composite Films with Regenerated Cellulose Prepared via Acid-Catalytic Degradation for in-Situ Growth of ZnO. *Int. J. Biol. Macromol.* **2025**, *305*, 140386. [CrossRef] [PubMed]
23. Pollard, Z.A.; Karod, M.; Schmitz, A.; Pian, B.; Barstow, B.; Goldfarb, J.L. ZnO Precursor's Ability to Catalyze Formation of Reactive Oxygen Species to Degrade Aqueous Organic Pollutants. *Chem. Eng. J.* **2024**, *480*, 147499. [CrossRef]

24. Wang, X.-Y.; Wang, Y.; Zhao, N.-D.; Zheng, M.; Guo, Y.-R.; Pan, Q.-J. Van Der Waals Enhanced Interfacial Interaction in Cellulose/Zinc Oxide Nanocomposite Coupled by Graphitic Carbon Nitride. *Carbohydr. Polym.* **2021**, *268*, 118235. [CrossRef] [PubMed]
25. Zafar, M.N.; Dar, Q.; Nawaz, F.; Zafar, M.N.; Iqbal, M.; Nazar, M.F. Effective Adsorptive Removal of Azo Dyes over Spherical ZnO Nanoparticles. *J. Mater. Res. Technol.* **2019**, *8*, 713–725. [CrossRef]
26. Li, X.; Zhang, L.; Wang, Z.; Wu, S.; Ma, J. Cellulose Controlled Zinc Oxide Nanoparticles with Adjustable Morphology and Their Photocatalytic Performances. *Carbohydr. Polym.* **2021**, *259*, 117752. [CrossRef]
27. Katepetch, C.; Rujiravanit, R.; Tamura, H. Formation of Nanocrystalline ZnO Particles into Bacterial Cellulose Pellicle by Ultrasonic-Assisted in Situ Synthesis. *Cellulose* **2013**, *20*, 1275–1292. [CrossRef]
28. Zhao, S.-W.; Zheng, M.; Zou, X.-H.; Guo, Y.; Pan, Q.-J. Self-Assembly of Hierarchically Structured Cellulose@ZnO Composite in Solid-Liquid Homogeneous Phase: Synthesis, DFT Calculations, and Enhanced Antibacterial Activities. *ACS Sustain. Chem. Eng.* **2017**, *5*, 6585–6596. [CrossRef]
29. Akpomie, K.G.; Adegoke, K.A.; Oyedotun, K.O.; Ighalo, J.O.; Amaku, J.F.; Olisah, C.; Adeola, A.O.; Iwuozor, K.O.; Conradie, J. Removal of Bromophenol Blue Dye from Water onto Biomass, Activated Carbon, Biochar, Polymer, Nanoparticle, and Composite Adsorbents. *Biomass Conv. Bioref.* **2024**, *14*, 13629–13657. [CrossRef]
30. Zhang, G.; Rong, X.; Huang, W.; Xiao, Z.; Xue, Z.; Cheng, H.; Feng, J.; Li, L. Preparation of Cellulose-Based Carbon Aerogel Composite Zinc Oxide Nanomaterials and Their Photocatalytic Properties. *J. Mol. Struct.* **2025**, *1327*, 141163. [CrossRef]
31. Sánchez-Gutiérrez, M.; Bascón-Villegas, I.; Espinosa, E.; Carrasco, E.; Pérez-Rodríguez, F.; Rodríguez, A. Cellulose Nanofibers from Olive Tree Pruning as Food Packaging Additive of a Biodegradable Film. *Foods* **2021**, *10*, 1584. [CrossRef] [PubMed]
32. Belhaj, J.; Serrano, L.; Khiari, R.; Garcia, A. The Cellulose Fibre Industry: Harnessing Agricultural Waste for Production. *Cellul. Chem. Technol.* **2024**, *58*, 929–936. [CrossRef]
33. Badawy, A.A.; Ghanem, A.F.; Yassin, M.A.; Youssef, A.M.; Abdel Rehim, M.H. Utilization and Characterization of Cellulose Nanocrystals Decorated with Silver and Zinc Oxide Nanoparticles for Removal of Lead Ion from Wastewater. *Environ. Nanotechnol. Monit. Manag.* **2021**, *16*, 100501. [CrossRef]
34. Sluiter, A.; Hames, B.; Ruiz, R.; Scarlata, C.; Sluiter, J.; Templeton, D. *Determination of Ash in Biomass*; National Renewable Energy Laboratory: Golden, CO, USA, 2007.
35. Beck, S.; Méthot, M.; Bouchard, J. General procedure for determining cellulose nanocrystal sulfate half-ester content by conductometric titration. *Cellulose* **2015**, *22*, 101–116. [CrossRef]
36. Liu, M.; Wan, Y.; Zhu, C.; Chen, G.; Li, X. FeNi bimetallic modified pg-C3N4-x and its photoelectrocatalytic hydrogen production coupling anodic oxidation. *Sep. Purif. Technol.* **2025**, *357*, 130142. [CrossRef]
37. Dehghani, M.; Nadeem, H.; Singh Raghuwanshi, V.; Mahdavi, H.; Banaszak Holl, M.M.; Batchelor, W. ZnO/Cellulose Nanofiber Composites for Sustainable Sunlight-Driven Dye Degradation. *ACS Appl. Nano Mater.* **2020**, *3*, 10284–10295. [CrossRef]
38. Ruiz-Flores, A.; García, A.; Pineda, A.; Brox, M.; Gersnoviez, A.; Cañete-Carmona, E. Low-Cost Photoreactor to Monitor Wastewater Pollutant Decomposition. *Sensors* **2023**, *23*, 775. [CrossRef]
39. Ezzine, M.; El-Shafie, A.S.; Youssef, K.M.; El-Azazy, M. Bifunctional TiO_2—Cellulose Based Nanocomposites for Synergistic Adsorptive-Photocatalytic Removal of Methyl Orange: Response Modelling and Optimization. *Int. J. Biol. Macromol.* **2025**, *306*, 141753. [CrossRef]
40. Li, H.; Zhang, L.; Lu, H.; Ma, J.; Zhou, X.; Wang, Z.; Yi, C. Macro-/Nanoporous Al-Doped ZnO/Cellulose Composites Based on Tunable Cellulose Fiber Sizes for Enhancing Photocatalytic Properties. *Carbohydr. Polym.* **2020**, *250*, 116873. [CrossRef]
41. García, A.; Labidi, J.; Belgacem, M.N.; Bras, J. The Nanocellulose Biorefinery: Woody versus Herbaceous Agricultural Wastes for NCC Production. *Cellulose* **2017**, *24*, 693–704. [CrossRef]
42. Kołodziejczak-Radzimska, A.; Markiewicz, E.; Jesionowski, T. Structural Characterisation of ZnO Particles Obtained by the Emulsion Precipitation Method. *J. Nanomater.* **2012**, *2012*, 656353. [CrossRef]
43. Bah, A.; Lim, K.Y.; Wei, F.; Khursheed, A.; Sow, C.H. Fluorescence Invigoration in Carbon-Incorporated Zinc Oxide Nanowires from Passage of Field Emission Electrons. *Sci. Rep.* **2019**, *9*, 9671. [CrossRef]
44. Cho, S.; Jang, J.-W.; Lee, J.S.; Lee, K.-H. Carbon-Doped ZnO Nanostructures Synthesized Using Vitamin C for Visible Light Photocatalysis. *CrystEngComm* **2010**, *12*, 3929. [CrossRef]
45. Isse, A.A.; Gennaro, A. Absolute Potential of the Standard Hydrogen Electrode and the Problem of Interconversion of Potentials in Different Solvents. *J. Phys. Chem. B* **2010**, *114*, 7894–7899. [CrossRef] [PubMed]
46. Rodwihok, C.; Charoensri, K.; Wongratanaphisan, D.; Choi, W.M.; Hur, S.H.; Park, H.J.; Chung, J.S. Improved Photocatalytic Activity of Surface Charge Functionalized ZnO Nanoparticles Using Aniline. *J. Mater. Sci. Technol.* **2021**, *76*, 1–10. [CrossRef]
47. Mahmoud, N.M.R.; Elbadawy, H.A.; Refaat, H.M. An Efficient Green Protocol for Photo-Degradation of Bromophenol Blue Dye. *Desalination Water Treat.* **2021**, *229*, 403–412. [CrossRef]

48. Akpomie, K.G.; Ghosh, S.; Gryzenhout, M.; Conradie, J. One-Pot Synthesis of Zinc Oxide Nanoparticles via Chemical Precipitation for Bromophenol Blue Adsorption and the Antifungal Activity against Filamentous Fungi. *Sci. Rep.* **2021**, *11*, 8305. [CrossRef]
49. Shah, T.; Gul, T.; Saeed, K. Photodegradation of Bromophenol Blue in Aqueous Medium Using Graphene Nanoplates-Supported TiO_2. *Appl. Water Sci.* **2019**, *9*, 105. [CrossRef]

polymers

MDPI

Article

Synthesis of Acidic Phosphonic Chitosan and the Complexation of La(III) in Acidic Aqueous Solution

Min Zhou †, Zhenglin Liu †, Dandan Lu, Jiajun Wang, Zili Chen and Yunren Qiu *

School of Chemistry and Chemical Engineering, Central South University, Changsha 410083, China; 222312163@csu.edu.cn (M.Z.); 232312186@csu.edu.cn (Z.L.); ludandan0707@163.com (D.L.); 242312177@csu.edu.cn (J.W.); zlchen0212@outlook.com (Z.C.)

* Correspondence: csu_tian@csu.edu.cn

† These authors contributed equally to this work.

Abstract: Due to the similar physicochemical properties of rare earth ions, their separation presents significant challenges. In this study, acidic phosphonic chitosan (aPCS) was prepared by modifying chitosan with phosphite and formaldehyde for improving the water solubility and complexing ability of rare earth ions in acidic aqueous solutions. DFT calculations revealed that its phosphonic groups serve as preferred reaction sites, forming stable bidentate complexes with rare earth cations. The complexation abilities of aPCS and phosphorylated chitosan (PCS) for La(III) were compared at various pHs by complexation–ultrafiltration. The results showed that aPCS achieved a 97% rejection for La(III), while 70% for PCS at pH 5 and P/RE 10. Furthermore, aPCS maintained higher rejection than PCS at pH of 3 to 7. In conclusion, aPCS demonstrates excellent potential for the selective extraction and purification of rare earth ions.

Keywords: rare earth separation; phosphorylated chitosan; complexation–ultrafiltration; acidic aqueous solution

1. Introduction

As a national strategic resource, rare earth elements have become crucial in many areas of contemporary society [1]. However, the physical and chemical properties of rare earth elements are strikingly similar, especially considering that their ionic radii change minimally with increasing atomic number [2–4]. This subtle variation makes the separation of these elements a formidable challenge. At present, different techniques have been innovated for separating rare earth elements including ion exchange, solvent extraction, membrane, ion imprinting [5]. Among these technologies, complexation–ultrafiltration (C–UF) is an effective way to separate rare earth elements. The process involves complexing rare earth ions with water-soluble polymers that have nitrogen, sulfur, or carbonyl functional groups [6]. These polymers form large molecular coordination compounds with the rare earth ions. An ultrafiltration membrane with a molecular weight cut-off higher than that of the coordination compounds is then used to retain these large complexes, while allowing other free ions to pass through. This method effectively reduces the concentration of rare earth ions or selectively separates them from the solution [7,8].

Complexing agents are integral to complexation ultrafiltration. The challenges posed by the similar chemical properties of rare earth elements and the limited selectivity of current agents have driven extensive research into the development of more efficient complexing agents. It has been observed that rare earth ions primarily form ionic bond complexes with most inorganic ligands. However, complexes formed with phosphorus ligands exhibit a covalent bond structure that resembles a chelate form, rendering them more stable than complexes with other inorganic ligands [9]. Amide extractants are particularly promising due to their ease of synthesis, resistance to hydrolysis, complete combustibility, and minimal environmental impact. When compared to amines and phosphorus extractants, amides possess an N-C=O functional group that allows for electron delocalization across the nitrogen, carbon, and oxygen atoms. This delocalization increases the electronegativity of the coordinating atoms, thereby enhancing their interaction with rare earth elements and improving the stability of the resulting complexes. Additionally, previous studies have demonstrated that acyl and amino-α-phosphamide compounds exhibit superior performance with rare earth ions through biocomplexation, showcasing both excellent selectivity and coordination capabilities for rare earth elements [10].

Chitosan is the second-most abundant natural organic polymer, after cellulose. Due to its non-toxic and harmless biological characteristics, chitosan has promising applications in high-tech fields such as medicine, cosmetics, environmental protection, and biomedical materials [11,12]. However, the adsorption capacity of chitosan for heavy metals is relatively low, and its molecular rigidity and solubility are poor, which greatly restricts its range of applications. Chitosan contains a large number of amino and hydroxyl groups, but due to steric hindrance and differences in charge, most amino groups are in a free state, exhibiting higher reactivity [13]. Consequently, despite the greater number of hydroxyl groups, modifications to chitosan typically occur at the amino sites.

By modifying chitosan, its physicochemical properties can be improved, resulting in modified chitosan materials with superior mechanical strength, good stability, high selectivity, and enhanced adsorption capacity [14,15]. This significantly broadens the application fields of chitosan and its derivatives. Additionally, modification of the amino groups in chitosan tends to improve its water solubility more effectively than modification at the hydroxyl sites [16]. For example, Tan et al. [17] modified hydroxypropyl chitosan to obtain N-acylated chitosan derivatives, demonstrating that the reactivity of amino groups in chitosan is higher and more easily acylated compared to hydroxyl groups. Zhang et al. [13] successfully synthesized an amino-modified tailings/chitosan composite, which was then applied as an adsorbent for the effective removal of Pb(II) and Cd(II) from aqueous solutions.

Although phosphonic chitosan derivatives have been previously studied for rare earth ion removal, most of these works primarily focus on their use as adsorbents [18,19]. Such adsorption processes are inherently restricted by the available contact area between the solid adsorbent and metal ions in solutions, which often results in lower efficiency. Reduced complexation efficiency or insufficient selectivity under acidic aqueous solution environments is another challenge, which is more relevant to practical extraction and separation processes. Addressing these challenges is crucial for advancing rare earth recovery technologies.

In our previous work, we synthesized phosphorylated chitosan (PCS), which exhibited superior rare earth ion complexation performance compared to polyacrylic acid sodium (PAAS) [20]. However, its complexation efficiency was limited in acidic aqueous solutions, a common environment in rare earth separation processes. To address this limitation, we developed an acidic phosphonic chitosan (aPCS) by modifying the amino groups of chitosan with phosphating agents. This modification not only enhanced the water solubility of chitosan but also significantly improved its complexation efficiency under acidic conditions. Characterization of aPCS using Fourier-transform infrared spectroscopy (FTIR), nuclear magnetic resonance spectra (NMR), X-ray diffraction (XRD), and scanning electron microscopy (SEM) confirmed its structural properties. C–UF experimental results demonstrated that aPCS achieved superior rejection rates for rare earth ions, surpassing the performance of PCS under challenging acidic conditions. This study aims to address the challenges of rare earth separation in acidic aqueous solutions by exploring the potential of aPCS as an efficient complexing agent.

2. Materials and Methods

2.1. Materials and Membrane

Chitosan with an average molecular weight of 300 kDa and a degree of deacetylation of 80%, as well as 99% methanesulfonic acid, were purchased from Shanghai Mai Lin Biochemical Co. (Shanghai, China). Phosphite was supplied by the Aladdin Chemical Reagent Company (Shanghai, China). Hydrochloric acid and sodium hydroxide were provided by Liaoning Pulide Chemical Technology Co. (Shenyang, China) Lanthanum oxide (La_2O_3), dehydrated alcohol, acetone and a 37–40% formaldehyde solution were sourced from Chengdu Cologne Chemical Co., Ltd. (Chengdu, China). MD44 dialysis bags with a retention range of 8–14 kDa (average diameter 28 mm) were purchased from Shanghai Leibsey Company (Shanghai, China). All reagents used were of analytical grade.

2.2. Methods

2.2.1. Synthesis of the aPCS

To improve the solubility and reactivity of chitosan, 5 g of chitosan was added to 25 g of sodium hydroxide dissolved in 75 mL of ultrapure water. The resulting mixture was stirred thoroughly at room temperature and stored in a refrigerator for 10 days. After thawing, the alkalized chitosan was collected by suction filtration, washed, and dried.

Weigh 0.5 g of the above alkalized chitosan dissolved in 150 mL of 5% methanesulfonic acid solution, stir thoroughly and transfer to a 500 mL round-bottomed flask, and then add 10 mL 40% phosphorous acid solution, stirring for 1 h to achieve full mixing. Increase the temperature to 75 °C, add an appropriate amount of 40% formaldehyde solution dropwise, and control the addition amount of phosphite and formaldehyde 1:1 (the molar ratio of amino groups in chitosan to phosphorous acid was approximately 1:17). The condensation and reflux reaction was then carried out for 7 h. After the reaction was completed, the solution was cooled to room temperature and dialyzed for 72 h (the MWCO of dialysis bag was 12,000 kDa). The dialysate was distilled under reduced pressure to obtain the product concentrate. A mixture of anhydrous ethanol and acetone was added to the concentrate, and a white flocculent precipitate was precipitated, which was allowed to stand overnight, filtered and washed, and then dried under vacuum at 55 °C to obtain the product, aPCS.

2.2.2. Procedure of C–UF

In this process, La(III) was selected as a representative of rare earth ions. The ultrafiltration membrane device used in this study was the polyethersulfone (PES) hollow fiber membrane, with a molecular weight cutoff (MWCO) of 20 kDa. Firstly, the complexes were pretreated using diafiltration to remove small polymers. Then, complexant solutions of polymers with rare earths were prepared with different P/RE values (mass ratio of complexant to rare earth ions). After approximately 2 h of exposure, the prepared material was introduced into hollow fiber and flat plate membrane modules using a peristaltic pump operating at a feed rate of 30 L/h. Permeate samples were collected for 30 s after a 5 min interval and subsequently analyzed to measure the concentrations of rare earth ions and polymers. All experiments were performed at 25 °C.

2.3. Characterization

FTIR analysis (Nicolet iS50, Thermo Fisher Scientific, Waltham, MA, USA) was used to identify the functional groups present in the raw materials and the final products. The final products, dissolved in D_2O, were analyzed using ^1H NMR and ^{31}P NMR spectroscopy (Bruker AVIII, 500 MHz, Zurich, Switzerland) to monitor and confirm the target compounds. The crystallinity of chitosan and aPCS was examined using XRD (Advance D8, Bruker AXS, Karlsruhe, Germany). The morphology of the products was observed by SEM on a JSM-7610F (JEOL, Tokyo, Japan) field emission scanning electron microscope. The concentration of La(III) in solution was determined using inductively coupled plasma optical emission spectrometry (ICP-OES, Avio 500, PerkinElmer, Singapore).

2.4. DFT Caculation

Structural optimization and single-point energy calculations of aPCS were performed using the B3LYP functional [21] with the 6-311G+** basis set [22]. The vibration frequencies are calculated after structural optimization. The DFT-D3 dispersion correction was applied to account for dispersion interactions, and the density-based implicit solvation model (SMD) was used to simulate the solvent environment. Then, the structures of aPCS–La $(H_2O)_6{}^+$ complexes was optimized using the PBE0 functional [23] and the def2-SVP basis set [24]. All calculations were carried out using the Gaussian 16(C01) software package [25]. The electrostatic potential (ESP) of the aPCS molecule was then calculated using the Multiwfn 3.8 (dev) code [26] to analyze its properties and identify the optimal reaction site. The isosurface plots in this work were all realized by rendering cube files exported from the Multiwfn 3.8 (dev) code using visual molecular dynamics (VMD 1.9.3) software [27].

3. Results and Discussion

Scheme 1 shows the synthesis mechanism of aPCS. The amine group of chitosan reacts with formaldehyde to form a positively charged iminium ion, which then undergoes electrophilic substitution with phosphorous acid. The resulting aPCS has a water solubility exceeding 2.9 mg mL^{-1}, significantly higher than that of chitosan.

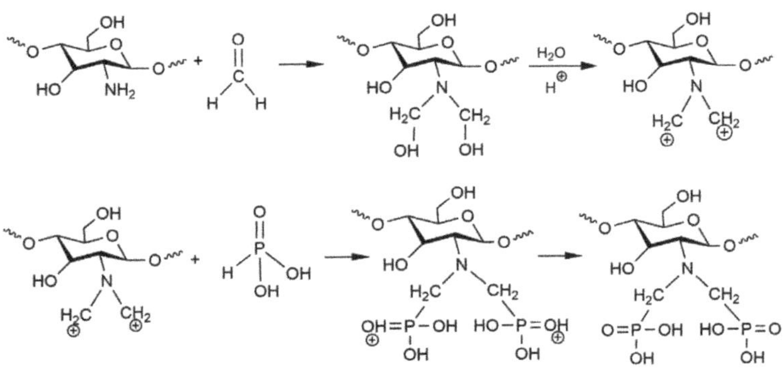

Scheme 1. Synthesis mechanism for the aPCS.

Figure 1 presents the FTIR spectra of the original chitosan and aPCS, CS presents a broad absorption band between 3464 and 3192 cm^{-1}, which is attributed to the stretching vibrations of N-H and O-H bonds [28]. The asymmetric and symmetric stretching vibration absorption peaks observed at 2933 cm^{-1} and 2870 cm^{-1} are due to the presence of C-H bonds in -CH_2- or -CH_3- groups, respectively. The absorption peaks at 1672 cm^{-1} and 1592 cm^{-1} correspond to the stretching vibrations of C-O bonds and the in-plane bending vibrations of N-H groups. In the case of aPCS, the amine methylation reaction grafts

-C-PO(OH)$_2$ onto the C2-NH$_2$ of CS, resulting in an intensified peak at 1471 cm^{-1}. This peak is characteristic of the stretching vibrations of phosphonate groups (-C-PO(OH)$_2$) and confirms the successful grafting of phosphonate moieties onto the molecular chain. Additionally, the broad absorption peak at 3464 cm^{-1} is retained but becomes even broader compared to CS, suggesting enhanced hydrogen bonding interactions. This broadening reflects the introduction of polar phosphonate groups, which increase the density and diversity of hydrogen bonding within the material, particularly in the non-crystalline regions. The hydroxyl absorption peak at 1370 cm^{-1} and the C-O bond absorption peak of the primary hydroxyl on C6 at 1021 cm^{-1} show minimal changes for aPCS, indicating that the modification primarily occurs at the C2-NH$_2$ group and leaves the hydroxyl groups at C6 largely unaffected. A new absorption peak at 2875 cm^{-1} appears in aPCS, corresponding to the C-H stretching vibration of methylene groups (-CH$_2$-), which confirms the involvement of formaldehyde in the amine methylation reaction. Furthermore, a strong peak at 1422 cm^{-1}, attributed to the stretching vibration of C-P=O bonds, is observed, further supporting the successful introduction of phosphonate groups [29]. These spectral changes collectively indicate that phosphite and methylene groups have been grafted onto the molecular chain of chitosan, primarily at the amine functional groups.

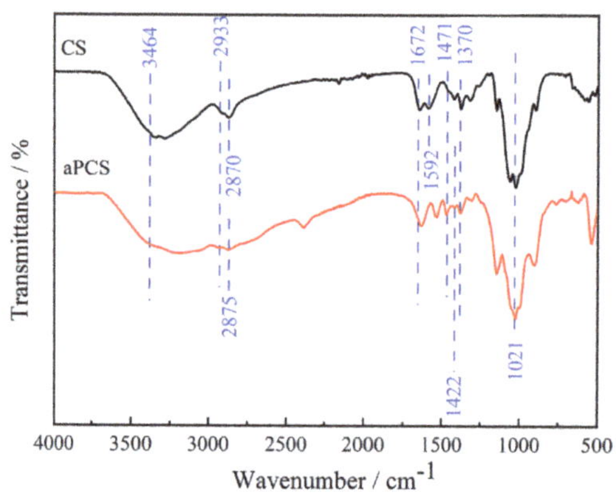

Figure 1. FTIR spectra of the aPCS and chitosan.

The ^{1}H-NMR and ^{31}P-NMR of aPCS are presented in Figure 2a,b, respectively. As shown in Figure 2a, the signal at 4.5–4.8 ppm corresponds to H1, the peaks between 3.4 and 4.3 ppm are assigned to the H3–H6 protons, and the signal at 3.14 ppm is attributed to H2. These assignments are consistent with the ^{1}H-NMR spectrum of chitosan reported in previous study [30]. For aPCS, the ^{1}H-NMR spectrum displays a distinct peak at 2.65–2.94 ppm, corresponding to the methylene between phosphorus and nitrogen atoms, indicating that the methylene groups are grafted onto the amino position. Additionally, a new peak at 6.19 ppm is attributed to -C-PO(OH)2, confirming the attachment of the phosphate groups to the methylene groups. Furthermore, the ^{31}P-NMR spectrum of aPCS, shown in Figure 2b, exhibits a prominent singlet at 2.56 ppm, providing direct evidence of phosphate groups incorporation.

The crystallinity of chitosan and aPCS is compared by XRD analysis, as shown in Figure 3. The pronounced diffraction peak at 20° for chitosan reflects its regular crystalline structure, driven by extensive hydrogen bonding. After modification, the diffraction intensity values found in aPCS decrease, indicating disruption of the crystalline structure of chitosan. This observation aligns with the FTIR results, where the newly introduced phosphonate groups weaken hydrogen bonding, reducing crystallinity and increasing molecular chain flexibility [31]. The increased molecular chain flexibility and reduced hydrogen bonding network improve the interaction of aPCS with water molecules, which directly contributes to its greater solubility in water compared to CS.

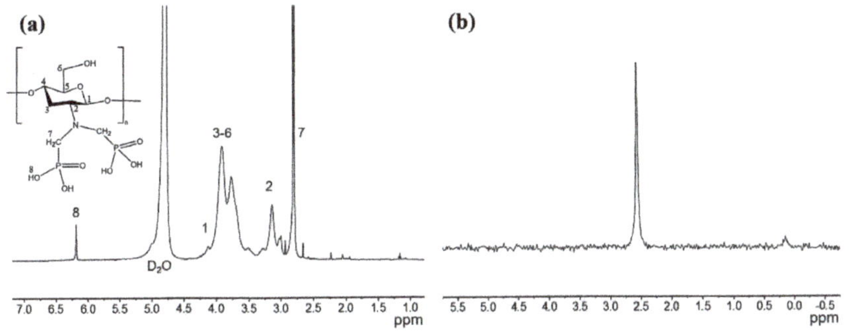

Figure 2. Schematic representation of the ^1H NMR (**a**) and the ^{31}P NMR spectrum of phosphonated chitosan (**b**).

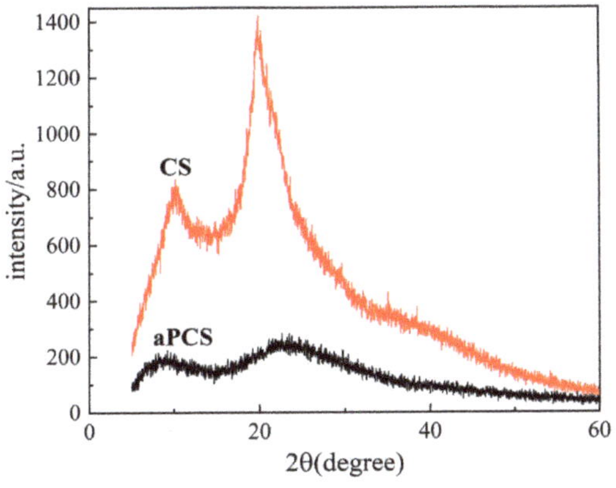

Figure 3. X-ray diffraction (XRD) patterns of the aPCS and CS.

Figure 4a,b presents the SEM images of raw materials of chitosan and aPCS at the same magnification. The CS exhibits a regular shape with a smooth surface, while the surface of aPCS is rough and porous, indicating the structural changes caused by the grafting of phosphonate groups. This disrupted crystalline structure correlates with the enhanced water solubility of aPCS. Concurrently, the element distribution of aPCS is analyzed by SEM–EDS mapping (Figure 4c,d). The product contains elements P, N, O, and C, with a higher quantity of P compared to N, the target material has been successfully synthesized.

Figure 5 presents the most stable structure of the aPCS molecule, with vibrational frequency analysis confirming the absence of imaginary frequencies, indicating that the optimized structure is the true global minimum on its potential energy surface. The figure also shows the natural population analysis (NPA) charge distribution of aPCS. Atomic charges provide a simple and intuitive way to describe charge distribution in a system. Atoms with stronger negative charges are more likely to attract positively charged nucleophiles, thereby facilitating chemical reactions [32]. In the aPCS molecule, the oxygen atoms O(1), O(2), O(8), and O(10) of the phosphonate groups exhibit the strongest affinity for rare earth cations. Electron delocalization within the phosphonate group leads to an equal distribution of negative charge across O(1) and O(2). However, the hydrogen bond between O(10) and H-O(11) weakens the electronegativity of O(8) and O(10) compared to O(1) and O(2), which are part of the same phosphonate group.

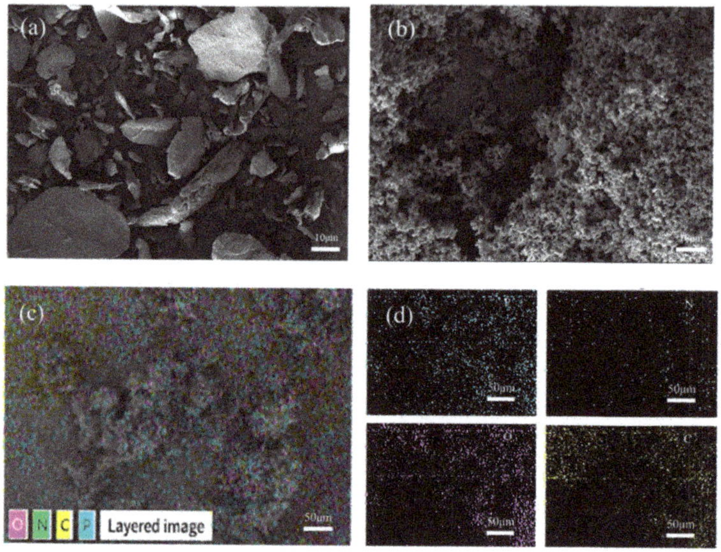

Figure 4. SEM images of CS (**a**) and aPCS (**b**), SEM-EDS mapping images of aPCS (**c**,**d**).

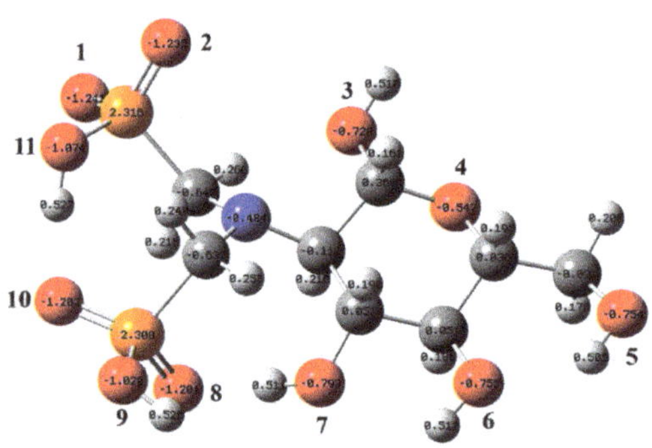

Figure 5. Optimized stable structure of aPCS and the NPA atomic charge distribution (white, gray, blue, red, and orange spheres represent H, C, N, O, and P atoms, respectively; only the oxygen atoms are labeled).

Given that the interaction between trivalent rare earth ions and organic ligands is predominantly ionic, ESP analysis is crucial for identifying the ligand's reactive sites and evaluating coordination strength. By analyzing the ESP on the ligand surface, regions likely to attract rare earth ions can be predicted [3]. Figure 6a shows the ESP distribution on the van der Waals surface of the aPCS molecule, with negative potential (blue) indicating areas susceptible to electrophilic attack, and positive potential (red) indicating nucleophilic attack sites. Cyan and yellow spheres represent the minima and maxima of the electrostatic potential, respectively, with the global minimum highlighted in magenta. The van der Waals surface of the phosphonate groups is predominantly deep blue or white, with no significant red regions, and the global minimum of the electrostatic potential (−9.82 eV) is located in this region. This suggests that this portion of the surface is electron-rich due to deprotonation, forming an anionic system that has a stronger attraction to rare earth cations. In our previous work, it was demonstrated that PCS primarily attracts rare earth ions through -NH and -P=O groups, subsequently forming coordination bonds [33].

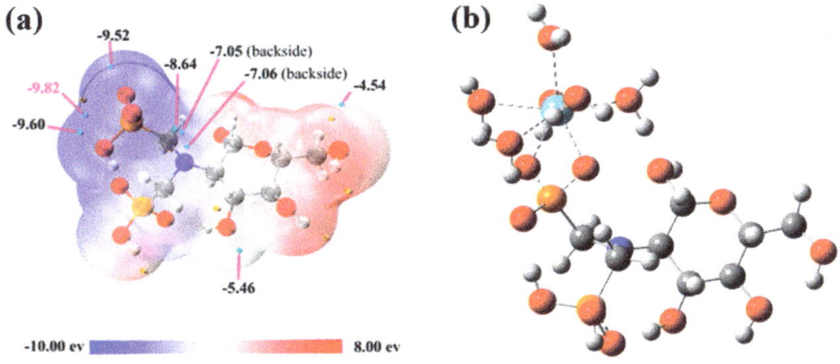

Figure 6. Distribution of ESP on the surface of aPCS molecules (**a**) and stable structure of aPCS–La $(H_2O)_6^+$ after optimization (cyan sphere represents La atom) (**b**).

Figure 6b shows the optimized structure of the complex after placing La(III) at the global minimum binding site of aPCS, visually illustrating the complexation mechanism between aPCS and La(III). The results are consistent with the ESP analysis, which suggests that the phosphonate groups in aPCS, upon deprotonation to form an anionic system, not only attract rare earth ions via long-range electrostatic interactions but also tend to coordinate in a bidentate manner with rare earth cations [34], thereby forming more stable complexes than previously synthesized PCS.

Building on the DFT analysis, which revealed favorable structural and electronic properties of aPCS for rare earth ion complexation, we experimentally investigated its ability to complex La(III).

Figure 7 shows the elemental distribution of the aPCS–La complex as analyzed by SEM–EDS mapping. The P element remains abundant, consistent with the structure of aPCS. Notably, La exhibits a spatial distribution that closely overlaps with that of P. This spatial overlap indicates that La(III) are coordinated to the phosphonate groups of aPCS, providing preliminary evidence for the formation of aPCS-La complexes.

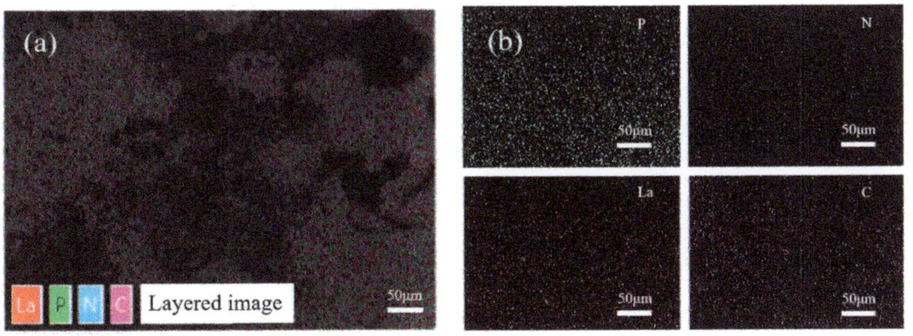

Figure 7. SEM-EDS mapping image showing the element distribution of aPCS-La complexes (**a**) and individual elemental mapping images corresponding to P, N, La, and C (**b**).

Ultrafiltration membranes were used to retain the aPCS-La complexes. The rejection rate (R), which represents the effect of complexation, was calculated as the ratio between the concentration of La(III) in the retentate and that in the original solution. As shown in Figure 8a, at pH 5 and P/RE 6, aPCS achieved a rejection rate exceeding 85% for La(III), compared to only 45% for PCS. When the P/RE ratio was increased to 10, the number of complexation sites provided by aPCS was sufficient to coordinate La(III) in solution. At this ratio, aPCS achieved a significantly higher rejection rate of 97%, compared to 70% for PCS. Furthermore, as shown in Figure 8b, the rejection rates of aPCS and PCS were evaluated at P/RE 10 across different pH, with aPCS consistently outperforming PCS. Additionally, the results of the ultrafiltration experiments provide strong evidence that La(III) ions are effectively complexed with both aPCS and PCS. This is demonstrated by the significantly lower concentration of rare earth ions in the permeate compared to the feed solution, indicating that the majority of the ions were retained as stable complexes. The superior performance of aPCS under acidic conditions can be attributed to the dissociation of hydrogen atoms from its phosphonate groups, forming an anionic system that enhances electrostatic interactions with La(III). The structural features of aPCS effectively promote the formation of stable complexes with La(III) at the acidic pH range of 3 to 7. These results underscore the potential of aPCS as an efficient ligand for rare earth recovery in industrial applications, particularly under acidic conditions.

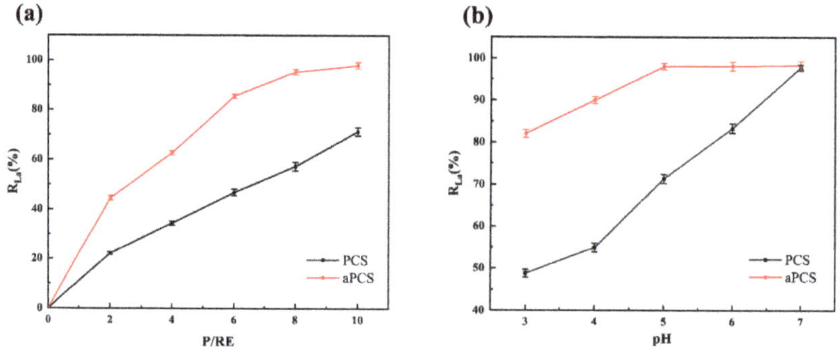

Figure 8. Comparison of complexing properties between aPCS and PCS at pH 5 (**a**), P/RE 10 (**b**).

4. Conclusions

In summary, aPCS has been successfully synthesized as a highly effective complexing agent for enhancing its complexation performance of rare earth ions in acidic aqueous solutions. The modification of chitosan with phosphonate groups significantly improves its water solubility and introduced functional sites capable of coordinating with rare earth elements. DFT analysis revealed that aPCS possesses favorable for bidentate coordination with La(III), indicating a strong binding affinity and structural stability of the resulting complexes. Experimental results confirmed that at pH 5, aPCS achieves a remarkable rejection rate of 97% for La(III), demonstrating its superior complexation performance in comparison to PCS. Moreover, aPCS maintains higher rejection rates than PCS within the acidic pH range of 3 to 7, emphasizing its adaptability to varying conditions. The preparation of aPCS is straightforward, making it a promising candidate for the selective separation and purification of rare earth ions in acidic aqueous solutions.

Author Contributions: Conceptualization, M.Z., D.L. and Z.C.; methodology, M.Z. and Y.Q.; software, M.Z.; validation, M.Z., Z.C. and D.L.; formal analysis, M.Z. and Z.L.; investigation, Z.L. and J.W.; resources, J.W.; data curation, M.Z. and Z.L.; writing—original draft preparation, M.Z. and Z.L.; writing—review and editing, D.L.; visualization, M.Z.; supervision, Y.Q.; project administration, Y.Q.; funding acquisition, Y.Q. All authors have read and agreed to the published version of the manuscript.

Funding: This work was supported by the National Natural Science Foundation of China (22178392) and the High Performance Computing Center of Central South University, China.

Institutional Review Board Statement: Not applicable.

Data Availability Statement: Data are contained within the article.

Conflicts of Interest: The authors declare no conflicts of interest.

References

1. Chen, Z.; Li, Z.; Chen, J.; Kallem, P.; Banat, F.; Qiu, H. Recent advances in selective separation technologies of rare earth elements: A review. *J. Environ. Chem. Eng.* **2022**, *10*, 107104. [CrossRef]
2. Liu, T.; Chen, J. Extraction and separation of heavy rare earth elements: A review. *Sep. Purif. Technol.* **2021**, *276*, 119263. [CrossRef]
3. Xu, L.; Pu, N.; Li, Y.; Wei, P.; Sun, T.; Xiao, C.; Chen, J.; Xu, C. Selective Separation and Complexation of Trivalent Actinide and Lanthanide by a Tetradentate Soft–Hard Donor Ligand: Solvent Extraction, Spectroscopy, and DFT Calculations. *Inorg. Chem.* **2019**, *58*, 4420–4430. [CrossRef]
4. Xiao, Y.; Feng, Z.; Huang, X.; Huang, L.; Chen, Y.; Liu, X.; Wang, L.; Long, Z. Recovery of rare earth from the ion-adsorption type rare earths ore: II. Compound leaching. *Hydrometallurgy* **2016**, *163*, 83–90. [CrossRef]
5. Traore, M.; Gong, A.; Wang, Y.; Qiu, L.; Bai, Y.; Zhao, W.; Liu, Y.; Chen, Y.; Liu, Y.; Wu, H.; et al. Research progress of rare earth separation methods and technologies. *J. Rare Earths* **2023**, *41*, 182–189. [CrossRef]
6. Kahloul, M.; Mahfoudhi, S.; Ounifi, I.; Elabed, B.; Amor, T.B.; Hafiane, A. Green complexation for heavy metals removal from wastewater by Keggin-polyoxometalates enhanced ultrafiltration. *Water Sci. Technol.* **2022**, *86*, 1510–1526. [CrossRef] [PubMed]
7. Li, J.; Wen, Y.; Chu, Y.; Liu, W.; Liu, J.; Wang, R.; Xu, Z. Separation of Rare Earth and Aluminum by Selective Complexation. *Minerals* **2022**, *12*, 743. [CrossRef]
8. Liu, L.; Liu, Y.; Luo, J.; Wan, Y.; Song, W. Membrane pre-concentration as an efficient strategy to enhance the enrichment of rare earth by MgO precipitation. *Chem. Eng. J.* **2024**, *491*, 152083. [CrossRef]
9. Kuang, S.; Zhang, Z.; Li, Y.; Wei, H.; Liao, W. Extraction and separation of heavy rare earths from chloride medium by α-aminophosphonic acid HEHAPP. *J. Rare Earths* **2018**, *36*, 304–310. [CrossRef]
10. Zhou, H.; Dong, Y.; Wang, Y.; Zhao, Z.; Xiao, Y.; Sun, X. Recovery of Th(IV) from leaching solutions of rare earth residues using a synergistic solvent extraction system consisting of Cyanex 572 and n-octyl diphenyl phosphate (ODP). *Hydrometallurgy* **2019**, *183*, 186–192. [CrossRef]
11. Pandit, A.H.; Nisar, S.; Imtiyaz, K.; Nadeem, M.; Mazumdar, N.; Rizvi, M.M.A.; Ahmad, S. Injectable, Self-Healing, and Biocompatible *N,O*-Carboxymethyl Chitosan/Multialdehyde Guar Gum Hydrogels for Sustained Anticancer Drug Delivery. *Biomacromolecules* **2021**, *22*, 3731–3745. [CrossRef]
12. Welchoff, M.A.; Wittenberg, A.T.; Jimenez, J.C.; Kamireddi, D.; Snelling, E.K.; Street, R.M.; Schauer, C.L. Post-modification of electrospun chitosan fibers. *Polym. Eng. Sci.* **2023**, *63*, 1921–1931. [CrossRef]
13. Zhang, Y.; Haris, M.; Zhang, L.; Zhang, C.; Wei, T.; Li, X.; Niu, Y.; Li, Y.; Guo, J.; Li, X. Amino-modified chitosan/gold tailings composite for selective and highly efficient removal of lead and cadmium from wastewater. *Chemosphere* **2022**, *308*, 136086. [CrossRef] [PubMed]
14. Ji, Z.; Zhang, Y.; Wang, H.; Li, C. Research progress in the removal of heavy metals by modified chitosan. *Tenside Surfactants Deterg.* **2022**, *59*, 281–293. [CrossRef]
15. Yin, Z.; Qiu, D.; Zhang, M. Molecular level study of cadmium adsorption on dithiocarbamate modified chitosan. *Environ. Pollut.* **2021**, *271*, 116322. [CrossRef]
16. Li, L.; Xu, Y.; Xu, Z.; Wu, C.; Chen, Q.; Xu, K.; Shi, Z. Synthesis, characterization and antifungal properties of dehydroabietic acid modified chitosan. *Int. J. Biol. Macromol.* **2024**, *255*, 128056. [CrossRef]
17. Tan, F.; Hao, N. Synthesis and Surface Property Characterization of N-Phthaloyl Hydroxypropyl Chitosan. *J. Wenshan Univ.* **2016**, *29*, 24–27.
18. Romal, J.R.A.; Ong, S.K. Opportunity for a greener recovery of dysprosium(III) from secondary sources by a novel Mannich reaction-modified phosphorylated chitosan hydrogel. *Int. J. Biol. Macromol.* **2024**, *267*, 131449. [CrossRef]
19. Hamza, M.F.; Abdellah, W.M.; Zaki, D.I.; Wei, Y.; Althumayri, K.; Brostow, W.; Hamad, N.A. A Phosphonic Functionalized Biopolymer for the Sorption of Lanthanum (III) and Application in the Recovery of Rare Earth Elements. *Sustainability* **2023**, *15*, 2843. [CrossRef]
20. Chen, Y.; Chen, Y.; Lu, D.; Qiu, Y. Synthesis of a Novel Water-Soluble Polymer Complexant Phosphorylated Chitosan for Rare Earth Complexation. *Polymers* **2022**, *14*, 419. [CrossRef]
21. Becke, A.D. Density-functional thermochemistry. III. The role of exact exchange. *J. Chem. Phys.* **1993**, *98*, 5648–5652. [CrossRef]
22. Baboul, A.G.; Curtiss, L.A.; Redfern, P.C.; Raghavachari, K. Gaussian-3 theory using density functional geometries and zero-point energies. *J. Chem. Phys.* **1999**, *110*, 7650–7657. [CrossRef]
23. Adamo, C.; Barone, V. Toward reliable density functional methods without adjustable parameters: The PBE0 model. *J. Chem. Phys.* **1999**, *110*, 6158–6170. [CrossRef]

24. Schäfer, A.; Horn, H.; Ahlrichs, R. Fully optimized contracted Gaussian basis sets for atoms Li to Kr. *J. Chem. Phys.* **1992**, *97*, 2571–2577. [CrossRef]
25. Frisch, M.J.; Trucks, G.W.; Schlegel, H.B.; Scuseria, G.E.; Robb, M.A.; Cheeseman, J.R.; Scalmani, G.; Barone, V.; Petersson, G.A.; Nakatsuji, H.; et al. *Gaussian 16 Rev. C.01*; Gaussian Inc.: Wallingford, CT, USA, 2016.
26. Lu, T.; Chen, F. Multiwfn: A multifunctional wavefunction analyzer. *J. Comput. Chem.* **2012**, *33*, 580–592. [CrossRef]
27. Humphrey, W.; Dalke, A.; Schulten, K. VMD: Visual molecular dynamics. *J. Mol. Graph.* **1996**, *14*, 33–38. [CrossRef] [PubMed]
28. Zidan, T.A.; Abdelhamid, A.E.; Zaki, E.G. N-Aminorhodanine modified chitosan hydrogel for antibacterial and copper ions removal from aqueous solutions. *Int. J. Biol. Macromol.* **2020**, *158*, 32–42. [CrossRef]
29. Liu, T.; Gou, S.; He, Y.; Fang, S.; Zhou, L.; Gou, G.; Liu, L. *N*-methylene phosphonic chitosan aerogels for efficient capture of Cu^{2+} and Pb^{2+} from aqueous environment. *Carbohydr. Polym.* **2021**, *269*, 118355. [CrossRef]
30. Sashiwa, H.; Kawasaki, N.; Nakayama, A.; Muraki, E.; Yamamoto, N.; Aiba, S.-i. Chemical Modification of Chitosan. 14: Synthesis of Water-Soluble Chitosan Derivatives by Simple Acetylation. *Biomacromolecules* **2002**, *3*, 1126–1128. [CrossRef]
31. Luo, X.; Zhao, L.; Khan, I.M.; Yue, L.; Zhang, Y.; Wang, Z. Chitosan-curcumin conjugate prepared by one-step free radical grafting: Characterization, and functional evaluation. *Carbohydr. Res.* **2024**, *545*, 109297. [CrossRef]
32. Wu, B.; Huang, J.; Tan, G.; Hao, Z.; Hu, G.; Luo, J.; Cui, C. DFT Study on the Structure and Properties of Hydantoin and Its Derivatives in the Alkaline Aqueous Solution. *Chemistry* **2021**, *84*, 610–619.
33. Li, W.; Zhou, M.; Fan, L.; Dai, Z.; Qiu, Y. Selective separation of Dy(III), Yb(III) and Y(III) by complexation and orderly decomplexation coupling with ultrafiltration based on rotating disk membrane and a DFT study. *Sep. Purif. Technol.* **2024**, *341*, 126665. [CrossRef]
34. Francos, J.; Elorriaga, D.; Crochet, P.; Cadierno, V. The chemistry of Group 8 metal complexes with phosphinous acids and related POH ligands. *Coord. Chem. Rev.* **2019**, *387*, 199–234. [CrossRef]

polymers

MDPI

Article

Preparation of Chitooligosaccharides with Specific Sequence Arrangement and Their Effect on Inducing Salt Resistance in Wheat Seedlings

Jingwen Li [1,2], Anbang Li [2], Yupeng Li [2], Siqi Zhu [2], Lin Song [1,*], Song Liu [2,3], Ronge Xing [2,3] and Kecheng Li [2,3,*]

1 College of Biological Engineering, Qingdao University of Science and Technology, Qingdao 266042, China; lijingwen223@163.com
2 CAS and Shandong Province Key Laboratory of Experimental Marine Biology, Institute of Oceanology, Chinese Academy of Sciences, Qingdao 266071, China; lianbang23@mails.ucas.ac.cn (A.L.); yupengli2023@163.com (Y.L.); 18255914694@163.com (S.Z.); sliu@qdio.ac.cn (S.L.); xingronge@qdio.ac.cn (R.X.)
3 Laboratory for Marine Drugs and Bioproducts, Qingdao Marine Science and Technology Center, Qingdao 266237, China
* Correspondence: 03231@qust.edu.cn (L.S.); lkc@qdio.ac.cn (K.L.); Tel.: +86-0532-82898512 (K.L.)

Abstract: Chitooligosaccharides (COS) exhibits good activity of inducing plant resistance, but the structure–activity relationship is still unclear. In this study, chitin oligosaccharides (CHOS) with a degree of polymerization (DP) of 2~6 were used as raw materials. Three deacetylases (NodB, *Vc*COD, and *Ar*CE4A) were employed to prepare three different sequence-arranged COSs, namely N-COS, C-COS, and A-COS, and their structures were characterized by infrared spectroscopy, high-performance liquid chromatography, and mass spectrometry. Further studies were conducted on inducing the plant salt resistance of the three different sequence-arranged COSs on wheat seedlings. The results showed a sequence-dependent effect of COS inducing plant salt resistance. Among them, A-COS exhibited the best activity. When sprayed at a concentration of 10 mg/L on wheat seedlings under salt stress for 3 days, the leaf length of the wheat seedlings sprayed with A-COS was recovered, and the wet mass and dry mass were recovered by 20.40% and 6.64%, respectively. Following the enhancement of proline accumulation, the malondialdehyde content decreased by 34.75%, and the Na^+/K^+ ratio also exhibited a significant reduction, thereby alleviating salt stress-induced damage. This study was the first to demonstrate the effect of COS with specific sequences on inducing plant salt resistance, providing a theoretical basis for the development of a new generation of efficient COS plant biostimulator.

Keywords: chitooligosaccharides; sequence arrangement; wheat; plant salt-resistance induction

1. Introduction

Salt stress is a common abiotic stress and is becoming an important factor affecting global food security [1,2]. Salt stress induces ionic toxicity, which disrupts the uptake of essential elements by plants, while the concomitant osmotic imbalance caused by high salinity leads to reduced stomatal conductance and a sharp decline in photosynthetic efficiency. At the same time, the oxidative stress induced by salt stress will produce a large amount of reactive oxygen species (ROS), causing lipid peroxidation and protein damage, ultimately leading to plant growth arrest or even death [3].

Current strategies to mitigate salt stress include exogenous chemical application, soil microenvironment optimization, agricultural water conservation technologies, and the cultivation of salt-tolerant cultivars [4]. Among these, exogenous chemical application is of particular importance due to its operational efficiency, targeted precision, and biocompatibility, working through direct interaction with stressed plants to activate antioxidant defense systems and enhance endogenous salt tolerance mechanisms. For example, the foliar application of salicylic acid (SA) effectively mitigates salt stress-induced malondialdehyde (MDA) accumulation in sunflower seedlings by activating the antioxidant defense system to scavenge ROS and attenuate lipid peroxidation-mediated cellular damage [5]. In addition, some natural polysaccharides also have good resistance-inducing effects. For example, the exogenous application of trehalose not only reduced MDA accumulation but also enhanced K^+ uptake capacity in salt-stressed tomato seedlings, ultimately improving their growth performance [6]. This indicates that the exogenous application of some natural polysaccharides is an effective treatment method to counteract the negative effects of salt stress.

Chitin, a biomass-derived natural polysaccharide, composed of $\beta(1\rightarrow4)$-glycosidic linked N-acetylglucosamine units, is the major component of the cell walls of various fungi and the exoskeletons of crustaceans and insects, serving as a critical biomolecule that triggers pathogen-associated molecular pattern (PAMP) responses through specific recognition by plant immune systems [7]. COS is the degradation product of deacetylated chitin and possesses good solubility, biocompatibility, and various biological activities such as antibacterial [8], the induction of chromium resistance [9], and salt tolerance [10]. COS has been reported to enhance salt tolerance in soybean seedlings through proline accumulation and lipid peroxidation mitigation under saline conditions, while simultaneously increasing shoot fresh biomass and promoting nutrient reserve deposition in plant tissues [11]. The induction of salt resistance in plants by COS has attracted extensive research attention, and it is a very promising plant biostimulator.

The molecular structure of COS determines its biological activity. COS is composed of alternating N-acetylglucosamine (**A**) and glucosamine (**D**), and the degree of acetylation (DA) is an important parameter affecting the biological activity of COS. Li et al. prepared COSs of three different DAs (0%, 32.2%, and 100%), and found that the fully deacetylated COS better promoted plant growth activity, which could promote the elongation of wheat roots and increase the stomatal conductance and increase the net photosynthetic rate and the biomass accumulation of wheat [12]. Using microwave-assisted KOH degradation, Fu et al. prepared COS with deacetylation degrees of 63.79%, 72.12%, 79.34%, and 88.15%, respectively. The results of the germination experiments on soaked wheat seeds showed that COS with a deacetylation degree of 72.12% could be the most effective in promoting the germination of wheat seeds, which was mainly achieved through increasing the activity of β-amylase and protease [13]. However, most COSs of different DAs were prepared by chemical methods in previous studies. The random distribution of monomers **A** and **D** in COS chains leads to the instability of the reported activity, and the structure–activity relationship between different acetyl sites and the biological activity of COS is not clear.

In recent years, chitin deacetylase has evoked increasing interest and could precisely remove one or more acetyl groups, thereby preparing COS with defined non-random acetylation sequences [14,15]. Zhu et al. prepared COSs using the deacetylase ScCDA2 from *Saccharomyces cerevisiae*, which selectively deacetylates the acetyl group at the non-terminal position of the sugar chain while retaining the acetyl group at the reducing end. The products were characterized by HPLC-ESI-MS and found to be seven COSs (**DAAA, ADAA, AADA, DDAA, DADA, ADDA**, and **DDDA**), thus providing an enzymatic strategy for the synthesis of specific COSs [16]. Hao et al. synthesized three chitobiose sequences (**AD, DA**, and **DD**) from an **AA** substrate using two deacetylases (NodB and *Vc*COD) and demonstrated that the different sequence architectures differentially modulate the antioxidative activity through the systematic evaluation of hydroxyl radical, superoxide radical, and DPPH scavenging activities coupled with reducing power assays [17]. Basa et al. also used these two deacetylases (NodB and *Vc*COD) in combination with chitinase and chitosanase to prepare several chitotetraoses with different sequences, **ADDD**, **DADD**, **DDDA**, and **DDDD**. It was found that **ADDD** and **DADD** showed strong stimulating activity on the oxidative burst of rice cells [18].

In summary, differences in the sequence arrangements of COSs lead to variations in their biological activity. However, the effect of different sequence arrangements of COS on its ability to induce salt resistance in plants remains unclear. In this study, a homogeneous sequence of chitin oligosaccharide (CHOS) was used as raw material, and three deacetylases with different deacetylation modes (NodB, *Vc*COD, and *Ar*CE4A) were employed to perform enzyme-catalyzed deacetylation. Three COSs with different sequence arrangements (N-COS, C-COS, and A-COS) were obtained. After the foliar application of low-concentration COS to wheat seedlings under salt stress, the physiological index, osmotic regulators content, MDA content, antioxidant enzyme activity, and Na^+/K^+ content were determined. The sequence effect on COS-induced wheat salt resistance was revealed, which provides crucial support for deciphering the sequence–activity relationship of COS in plant resistance induction. The clarification of the sequence arrangement of COS and its ability to induce salt resistance will help the development of the COS series of efficient antagonists, and the combination of COS with modern agricultural technologies such as nano-sized objects will help it play a better role in the field of agricultural production [19].

2. Materials and Methods

2.1. Materials

Chitin oligosaccharides with a degree of polymerization (DP) of 1~6 were purchased from the Tokyo Chemical Industry Co., Ltd. (Tokyo, Japan). The NodB sequence (NCBI acc. No. AJW76244.1) and *Vc*COD sequence (NCBI acc. No. AAF94439.1) were sourced from the NCBI database and provided to Sangon Biotech (Shanghai, China) Co., Ltd. for gene optimization and synthesis. And *Ar*CE4A was improved according to the method of Tuveng et al. [20], and its sequence was provided to GenScript Biotech Co. (Nanjing, China) for gene optimization and synthesis. *E. coli* BL21 (DE3) competent cells were purchased from Sangon Biotech (Shanghai, China) Co., Ltd. and used to express chitin deacetylases. A Malondialdehyde (MDA) Content Assay Kit, Peroxidase (POD) Activity Assay Kit, and Catalase (CAT) Activity Assay Kit were bought from Solarbio (Beijing, China). A Superoxide Dismutase (SOD) assay kit (WST-1 method) was bought from Nanjing Jiancheng Bioengineering Institute (Nanjing, China). All other chemicals and reagents were of analytical grade.

2.2. Preparation of COSs with Chitin Deacetylases

2.2.1. Heterologous Expression and Purification of Chitin Deacetylases

E. coli BL21 (DE3) was transformed into NodB, *Vc*COD, or *Ar*CE4A plasmids, respectively, and then grown in LB medium containing 1 μg/mL of ampicillin sodium (Amp) at 37 °C. When the OD600 reached 0.6~0.8, 0.1 mM isopropyl-β-D-thiogalactopyranoside (IPTG) was added to the culture medium for induction, and *E. coli* was cultured overnight at 16 °C. The cells were disrupted by sonication in Banding Buffer (50 mM Tris, 150 mM NaCl, pH = 8.0) and centrifuged at 4 °C to remove cell debris. Then, the protein was collected by gradient elution with 20~250 mM imidazole eluent in a Ni-agarose gel column. The collected eluent was analyzed by 12% SDS-PAGE, and the protein concentration was determined by the Coomassie brilliant blue dye method [21].

2.2.2. Preparation of COSs with Specific Sequence Arrangement

The monosaccharide in commercial chitin oligosaccharides was removed by alcohol precipitation to obtain CHOS with DPs of 2~6. NodB reacted with CHOS at a mass ratio of 1:20 overnight in a buffer (20 mM Mops, 10 mM DTT, 1 mM $MnSO_4$, pH = 7.5), and the product was purified by dialysis to obtain COS with the first deacetylation at the non-reducing-end residue (N-COS). *Vc*COD reacted with CHOS at a mass ratio of 1:20 overnight in a buffer (10 mM NH_4HCO_3, pH = 7.5), and the product was purified by dialysis to obtain COS with the deacetylation at the second residue from the non-reducing end (C-COS). The reaction of *Ar*CE4A with CHOS was carried out according to the method recorded by Tuveng et al. [21], and the reaction time was 48 h. After the reaction, the product was purified by dialysis to obtain multi-site deacetylated COS (A-COS).

2.3. Characterization

Powdered solid samples were detected in the range of 4000–500 cm^{-1} using an FTIR spectrometer, Thermo Scientific Nicolet iS10 (Thermo Fisher Scientific, Waltham, MA, USA), in ATR mode.

The LC-2030C 3D Plus HPLC system (SHIMADZU, Kyoto, Japan), equipped with the evaporative light detector (Essentia ELSD-16), was used for the analysis of COS by hydrophilic interaction liquid chromatography (HILIC). The chromatographic column was SHIMADZU ShimNex HE Amide (5 μm, 4.6 mm × 150 mm). The mobile phase was a binary mobile phase consisting of 20% ammonium formate solution and 80% acetonitrile. The flow rate was 1 mL/min, and the column temperature was 25 °C. The sample was filtered through a 0.22 nm needle filter prior to analysis.

ESI-MS was performed using LTQ Orbitrap XL (Thermo Fisher Scientific, Waltham, MA, USA). The ESI source voltage was 4 kV, the sheath gas was 10 arbs, the scan gas was 0 arbs, and the capillary temperature was 275 °C.

2.4. Plant Materials and Treatment

Wheat seeds (Jimai 22) were germinated in the dark at 25 °C for 24 h and then transferred to a culture cup. The temperature was maintained at 25/20 °C, the day/night cycle was 14/10 h, and the humidity was maintained at 65~70%. The Hoagland nutrient solution was replaced every other day. When the second leaf was fully developed, the wheat seedlings were randomly divided into seven experimental groups. Each experimental group had three parallels, including a control group (CK, no NaCl was applied, but distilled water was sprayed with 0.01% Tween-20), a negative control group (NaCl, 100 mM NaCl was applied, and distilled water was sprayed with 0.01% Tween-20), and a positive control group (SA, 100 mM NaCl was applied, and 100 mg/L of salicylic acid was sprayed with 0.01% Tween-20). The concentration of NaCl was in reference to Kashif et al. [22]. And the four sample groups were the CHOS group, N-COS group, C-COS group, and A-COS group. Each parallel group had 30 wheat seedlings. In total, 100 mM salt stress was applied to the culture solutions of the four groups of wheat, and the corresponding samples were sprayed at a concentration of 10 mg/L (0.01% of Tween-20 was added to the solution). Each parallel group was sprayed with 30 mL of sample solution each time, and the spraying process lasted for three days. After 10 days of treatment, the corresponding physiological and biochemical indexes of wheat seedlings were determined.

2.5. Determination of the Growth Parameters

After 10 days of treatment, the leaf length, root length, and fresh mass of fresh wheat seedlings were measured, and the dry mass was measured by baking wheat seedlings at 105 °C to a constant weight.

2.6. Determination of the Proline Content

After spraying the leaves with drugs for 10 days, wheat leaves were taken at the same position in the mortar, and 3% sulfosalicylic acid solution was added to the ice bath to extract proline. The free proline content was determined by the ninhydrin method [23], and the absorbance value of the microplate reader at 520 nm was recorded. The standard curve was drawn with L-proline as the standard to quantify the concentration.

2.7. Determination of the MDA Content

MDA is a product of lipid peroxidation in plants. The MDA content is used to determine the level of lipid peroxidation in plants. The same parts of wheat leaves were frozen and ground in liquid nitrogen, and the MDA content was determined using the MDA Content Kit.

2.8. Determination of Antioxidant Enzyme Activities

After 10 days of treatment, the same parts of wheat leaves were frozen in liquid nitrogen and homogenized to extract crude enzymes. The activities of T-SOD, POD, and CAT in wheat seedlings were determined by the T-SOD Activity Assay Kit, POD Activity Assay Kit, and CAT Activity Assay Kit, respectively.

2.9. Determination of Na^+ and K^+ Contents

After 10 days of treatment, the same parts of the wheat leaves were frozen in liquid nitrogen and ground. The contents of Na^+ and K^+ in wheat leaves were determined by Thermo ICPOES7200 (Thermo Fisher Scientific, Waltham, MA, USA). The emission power was 1.15 kW, the carrier gas was argon, the ion gas flow rate was 15 L/min, the auxiliary gas flow rate was 15 L/min, the nebulizer flow rate was 0.75 L/min, the detection mode was axial observation, and the calibration type was linear.

2.10. Statistical Analysis

All experimental data were analyzed by ANOVA using SPSS software (version 27.0), and the significant difference was detected by Duncan multiple range test ($p < 0.05$). Each dataset is expressed as the mean $\pm$ standard deviation of three independent replicates. Different letters (a, b, c...) on the same parameter indicate significant differences at the $p < 0.05$ level.

3. Results

3.1. Heterologous Expression and Purification of Deacetylase

To prepare accurate and non-random sequence COS, the deacetylases selected in this study for single-site deacetylation are NodB from *Sinorhizobium meliloti* and *Vc*COD from *Vibrio cholerae*. NodB removes the acetyl group at the first residue from the non-reducing end, while *Vc*COD removes the acetyl group at the second residue from the non-reducing end [24,25]. Since NodB and *Vc*COD deacetylate the acetyl groups on the non-reducing end, to clarify the specific effect of acetylation sites on the activity of COS, this study also selected a multi-site deacetylase (*Ar*CE4A) from *Marine Arthrobacter*. *Ar*CE4A preferentially removes the acetyl groups on the internal sugar chain, then removes the acetyl group at the first residue from the non-reducing end, and finally retains the acetyl group at the reducing end. As shown in Figure 1, NodB, *Vc*COD, and *Ar*CE4A are successfully expressed in *E. coli*. After purification using Ni-sepharose, SDS-PAGE analysis shows that the apparent molecular weights of NodB, *Vc*COD, and *Ar*CE4A are close to the expected 25 kDa, 48 kDa, and 29 kDa, respectively.

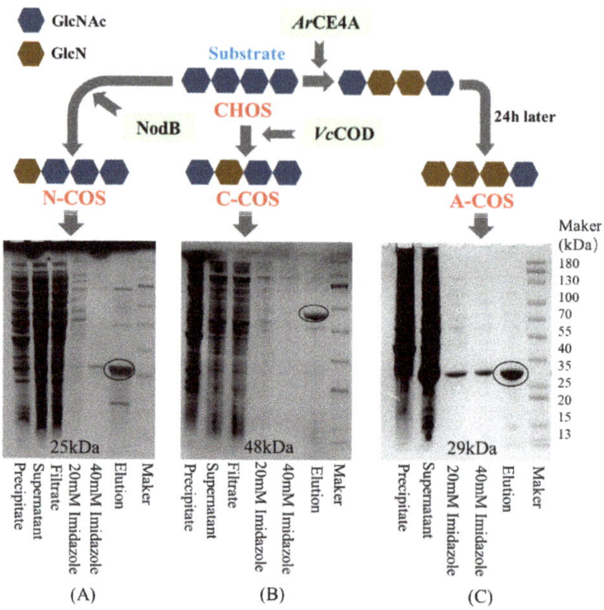

Figure 1. The results of the heterologous expression and purification of deacetylase and the deacetylation action sites (taking chitotetraose as an example). (**A**) SDS-PAGE of NodB; (**B**) SDS-PAGE of *Vc*COD; (**C**) SDS-PAGE of *Ar*CE4A.

3.2. Characterization of COSs with Specific Sequence Arrangements

The FTIR spectra of CHOS, N-COS, C-COS, and A-COS were analyzed, as shown in Figure 2. The characteristic peaks at 3275, 1635, 1546, and 1026 cm^{-1} correspond to the stretching vibration of −OH, the stretching vibration of C=O, the bending vibration of N-H, and the vibration of sugar ring in CHOS, respectively. CHOS exposes the amine groups after removing the acetyl groups, and the signal peak at 3200–3400 cm^{-1} changes from sharp to smooth due to the overlap of the stretching vibration of −OH and $-NH_2$. N-COS, C-COS, and A-COS showed this change after deacetylation, proving that the acetyl groups were successfully removed. However, due to the three types of COSs only removing a small amount of acetyl groups, the basic structure is still similar to the substrate CHOS, and the bending vibration peak at 1597 cm^{-1} corresponding to the free amine group in the full deacetylated COS infrared spectrum is not observed [26].

The HPLC analyses of CHOS, N-COS, C-COS, and A-COS were further performed, as shown in Figure 3. Since the HPLC uses a hydrophilic interaction system, the retention time of the components with lower molecular weight or smaller degrees of polymerization (DP) in CHOS is short. Therefore, the oligomers in CHOS are attributed to be chitin oligosaccharides with DP2, DP3, DP4, DP5, and DP6, respectively. After deacetylation, the amine groups exposed by deacetylase have better hydrophilicity than the acetylamino groups, and the affinity with the hydrophilic column is stronger, so the retention time is delayed. The retention time of N-COS and C-COS as mono-deacetylated COS is significantly delayed compared to CHOS, indicating that the two deacetylases have good activity and no substrate residue. The retention times of DPs of 2~6 after NodB and *Vc*COD deacetylation are also different, indicating that the two produced COSs are isomers. It also shows that the NodB and *Vc*COD expressed in this study have very good specificity and can specifically remove acetyl groups from the first and second residues at the non-reducing end, respectively. As a multi-site deacetylated COS, the retention delay of A-COS is more obvious. Compared with the removal of one acetyl group from N-COS and C-COS, the retention time of all components is less than 16 min, while A-COS still has components after 16 min, which also proves that the enzymatic reaction of *Ar*CE4A removes more acetyl groups from CHOS.

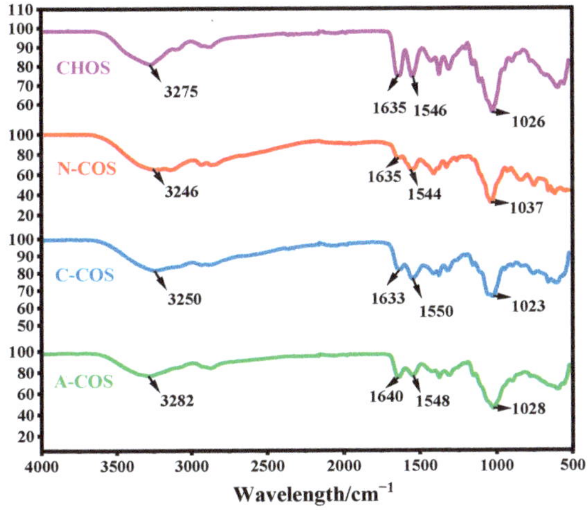

Figure 2. The FTIR spectra of CHOS, N-COS, C-COS, and A-COS.

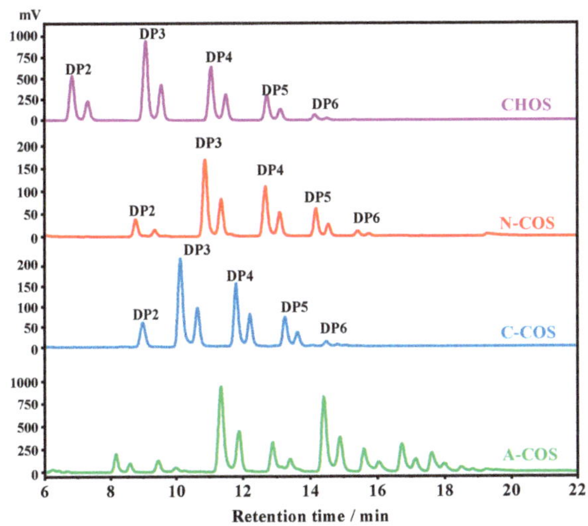

Figure 3. The HPLC spectra of CHOS, N-COS, C-COS, and A-COS.

The ESI-MS spectra of CHOS, N-COS, C-COS, and A-COS were carried out for the exact composition analysis. As shown in Figure 4A, the ion peaks of the main components are labeled as $[M + H]^+$ ion peaks corresponding to **AA**, **AAA**, **AAAA**, and **AAAAA**, respectively. Taking the ion peak with a m/z of 425.18 as an example [27], the calculation method is $425 = 2 \times 221$(GlcNAc) $- 18(H_2O) + 1$(H). Figure 4B shows the main components of N-COS (m/z 383), with one acetyl group removed, which are attributed to **DA**, **DAA**, **DAAA**, and **DAAAA**. N-COS and C-COS are isomers, both being mono-deacetylated COSs, so they show the same m/z signals in the ESI-MS spectra (Figure 4C). According to the deacetylation site of the *Vc*COD, C-COS mainly contains **AD**, **ADA**, **ADAA**, and

ADAAA. The ion peaks of **DA, D_2A, DA_2, D_3A, D_2A_2, D_4A,** and **AD_3A** are shown in Figure 4D. Based on the characteristics of *Ar*CE4A, the chitin deacetylase preferentially removes internal acetyl groups but has no activity on reducing terminal residues. Therefore, the products in A-COS are identified as **DA, DDA, ADA, DDDA, ADDA, DDDDA,** and **ADDDA**.

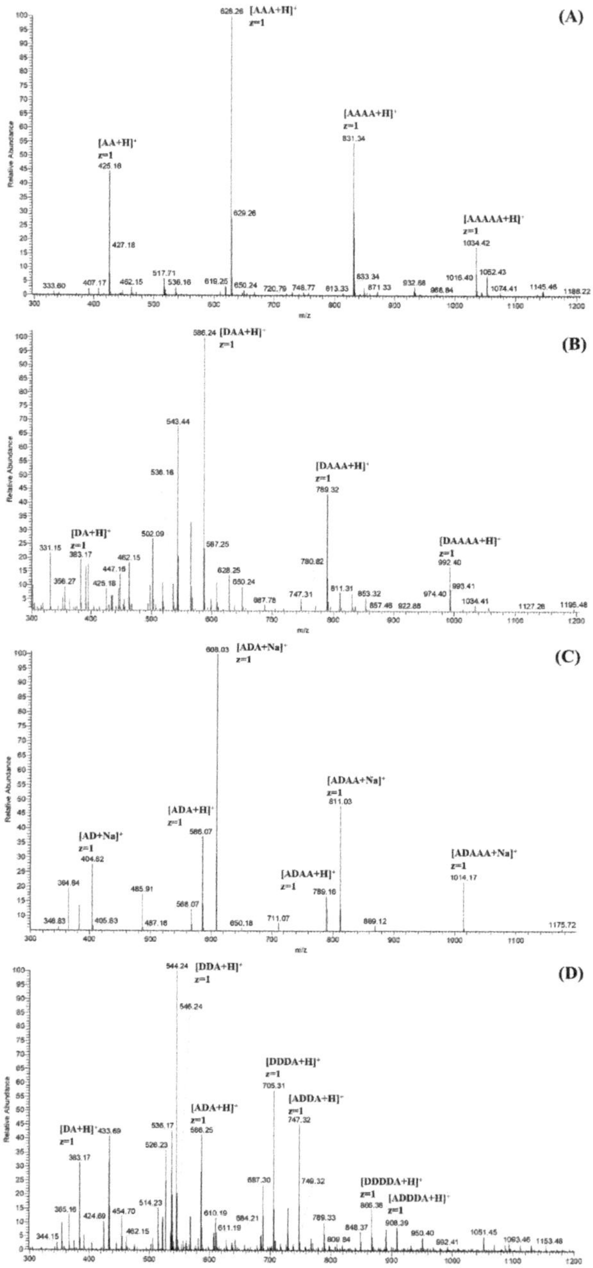

Figure 4. The ESI-MS spectra of CHOS (**A**), N-COS (**B**), C-COS (**C**), and A-COS (**D**).

3.3. Wheat Growth Index

As shown in Figure 5, 100 mM salt stress significantly reduced both fresh mass and dry mass in wheat seedlings. The foliar application of CHOS ($p = 0.007$, <0.05), C-COS ($p = 0.001$, <0.05), and A-COS ($p < 0.001$) significantly enhances fresh mass accumulation in wheat seedlings, with A-COS exhibiting the most pronounced efficacy by restoring fresh mass to the CK group. Salt stress exerts more pronounced impacts on dry mass, reducing dry mass to 59.73% of control levels, while CHOS ($p = 0.002$, <0.05) and A-COS ($p = 0.045$, <0.050) treatments significantly ameliorate this suppression compared with the NaCl group. Notably, the dry mass restorations of CHOS and A-COS formulations achieve comparable activity to the positive control group (salicylic acid, SA). Changes in the leaf length and root length of wheat seedlings under salt stress are not obvious. In summary, all growth indexes of wheat seedlings sprayed with A-COS under salt stress can be maintained at good levels.

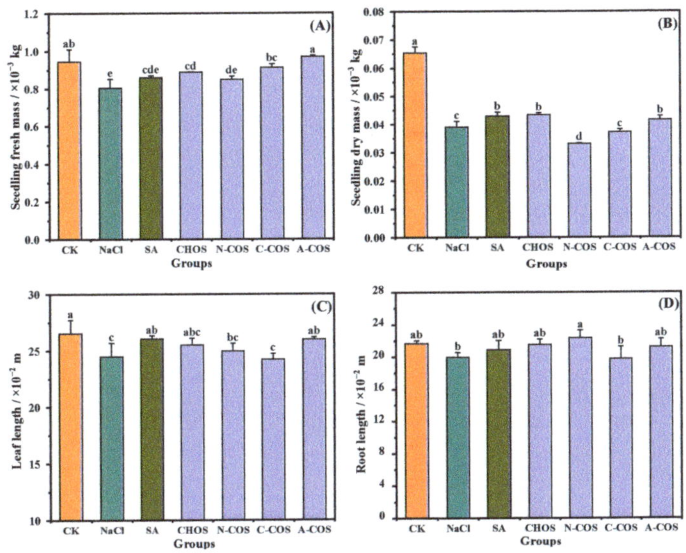

Figure 5. Effects of different treatment groups on growth indexes. (**A**) Seedling fresh mass; (**B**) seedling dry mass; (**C**) leaf length; (**D**) root length. Different letters (a,b,c,d,e) indicate significant differences at $p < 0.05$ on the same parameter.

3.4. Proline Content

Proline emerges as a critical multifunctional osmolyte in plant salt stress responses, serving a dual role as an indicator of ionic stress and a protective solute that stabilizes protein biosynthesis and metabolic processes. And it effectively prevents protein aggregation, the level of which is a key phenotypic indicator of superior salt tolerance in wheat varieties [28]. As shown in Figure 6, salt stress triggers a 90.34% increase in proline accumulation compared to control conditions, demonstrating the critical role of this compatible solute in the osmotic adjustment of wheat seedlings under saline conditions. After the SA application, the proline content of wheat seedlings is further increased, and the proline content of wheat seedlings in the SA group is 2.58 times higher than that in the CK group. A large gap appears in the changes in proline content after COS spraying, in which A-COS has an extremely significant promoting effect on the accumulation of proline in wheat seedlings, rising to 5.28 times that of the CK group.

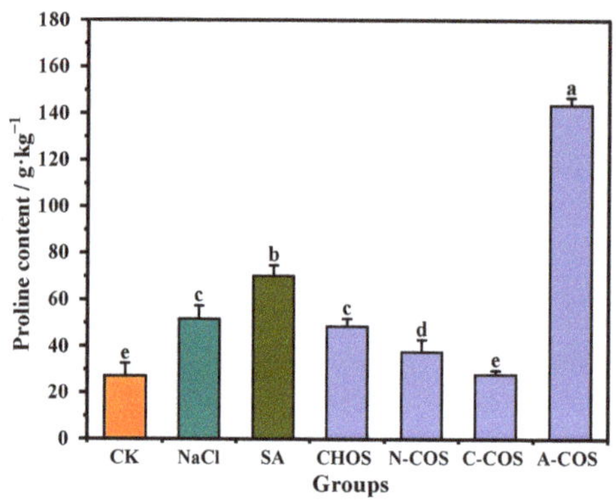

Figure 6. Effect of different treatment groups on proline content. Different letters (a,b,c,d,e) indicate significant differences at $p < 0.05$ on the same parameter.

3.5. MDA Content

Salt stress induces ROS overaccumulation in plants, which subsequently attack membrane-bound unsaturated fatty acids, thereby initiating lipid peroxidation cascades that generate cytotoxic metabolites including MDA and compromise membrane integrity. This oxidative assault disrupts cellular homeostasis by impairing developmental processes and metabolic functions, with MDA serving as a critical biomarker of lipid peroxidation. Reducing the accumulation of MDA in the plant can effectively alleviate oxidative damage in the plant [29]. The MDA contents of wheat seedlings in different treatment groups are shown in Figure 7. Salt stress results in increased lipid peroxidation in wheat seedlings, and the MDA content in the salt stress group is elevated by 80.64% compared with that in the CK group. The exogenous spraying of SA significantly alleviates this condition, and the MDA content is only 28.51% higher than that of the CK group. The foliar application of distinct COSs differentially modulates salt stress-induced oxidative damage in wheat seedlings. The treatment of CHOS maintains MDA levels comparable to the salt-stressed group, indicating a limited capacity to mitigate lipid peroxidation. N-COS, C-COS, and A-COS applications significantly reduce MDA levels by 20.51%, 25.47%, and 34.75%, respectively. Notably, A-COS demonstrates SA-equivalent efficacy in ameliorating lipid peroxidation, establishing it as a potent antioxidant intervention against salt-induced membrane damage.

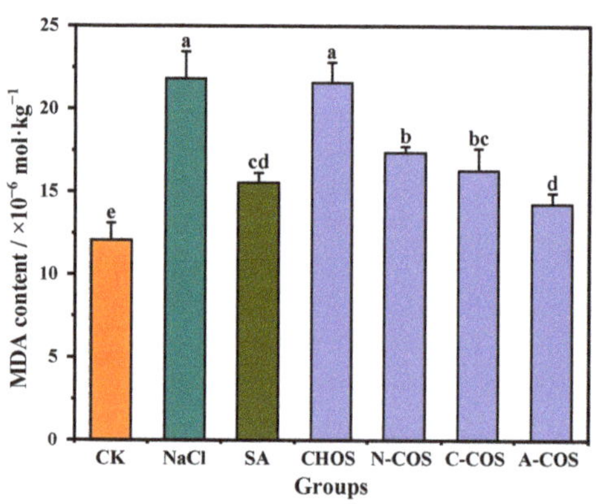

Figure 7. Effects of different treatment groups on MDA content. Different letters (a,b,c,d,e) indicate significant differences at $p < 0.05$ on the same parameter.

3.6. Antioxidant Enzyme Activities

ROS produced by plants under salt stress are scavenged by plant antioxidant enzymes, which are important components of oxidative defense in plants. POD, SOD, and CAT, as the main antioxidant enzymes, effectively reflect the stress response of the oxidative defense system to salt stress in plants [30]. SOD converts superoxide to hydrogen peroxide, CAT and POD function by scavenging hydrogen peroxide, and an effective dynamic equilibrium will be maintained between the activities of antioxidant enzymes and ROS in plant cells. As shown in Figure 8, POD (p = 0.004, <0.05) and CAT (p < 0.001) are activated under salt stress, showing an increase in enzyme activities. The foliar application of SA causes the activities of POD (p < 0.001) and CAT (p = 0.004, <0.05) to exhibit different degrees of regression. For wheat seedlings in the foliar-sprayed sample group, distinct samples show different effects. POD (p < 0.001), SOD (p = 0.002, <0.05), and CAT (p < 0.001) show an increase in activity in the presence of CHOS, while in the presence of A-COS, they all show a regression in activity, and the magnitude of this regression is consistent with that of SA.

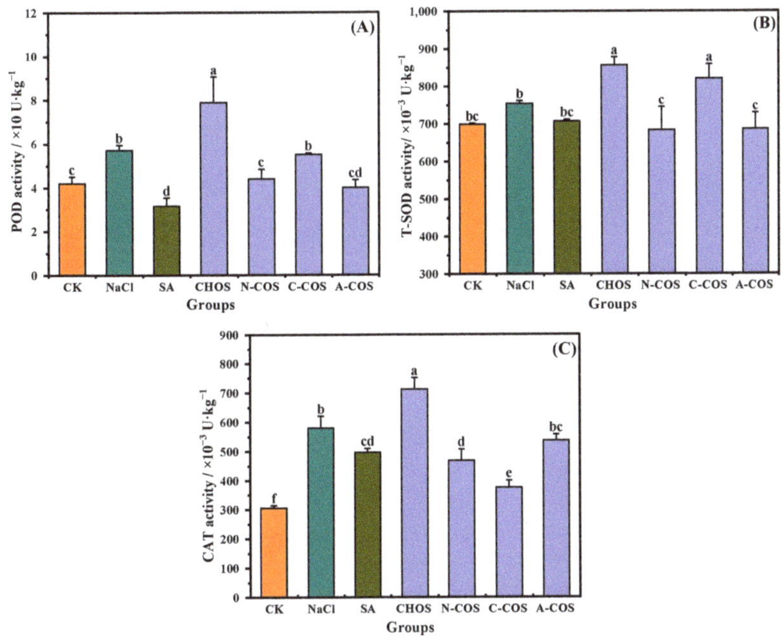

Figure 8. Effects of different treatment groups on antioxidant enzyme activities. (**A**) POD activity; (**B**) T-SOD activity; (**C**) CAT activity. Different letters (a,b,c,d,e,f) indicate significant differences at p < 0.05 on the same parameter.

3.7. Na^+ Content and K^+ Content

Salt stress-induced Na^+ hyperaccumulation in plant cells disrupts ionic homeostasis through competitive uptake via non-selective K^+ channels and high-affinity transporter proteins in root systems. This Na^+/K^+ antagonism critically impairs the uptake of K^+, which is an essential regulator for maintaining cellular osmotic equilibrium and sustaining photosynthetic machinery functionality [31]. As shown in Figure 9, salt stress induces a significant imbalance in the ionic homeostasis of wheat seedlings, a marked Na^+/K^+ ratio (p < 0.001), through enhanced Na^+ influx and suppressed K^+ assimilation. The foliar application of CHOS and COS significantly attenuates this ratio, with the hierarchy of efficacy being A-COS > N-COS > C-COS > CHOS. Notably, A-COS exhibits superior K^+

uptake enhancement, thereby mitigating Na^+-induced cytotoxicity and restoring membrane potential stability.

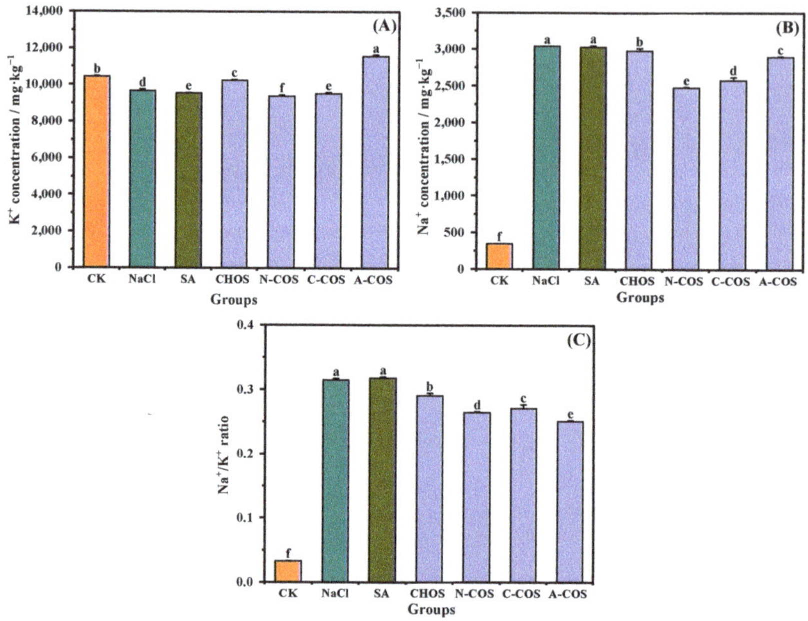

Figure 9. Effects of different treatment groups on Na^+ and K^+ contents. (**A**) K^+ concentration; (**B**) Na^+ concentration; (**C**) Na^+/K^+ ratio. Different letters (a,b,c,d,e,f) indicate significant differences at $p < 0.05$ on the same parameter.

4. Discussion

Salt stress impacts plant growth via diverse physiological mechanisms. The application of A-COS significantly improves multiple physiological parameters in wheat seedlings, completely restoring shoot fresh weight to normal levels. Wheat seedlings exhibit high sensitivity to salt stress, especially during the early stages of growth. Under salt stress, plants initiate metabolic regulation such as organic matter conversion for osmoregulation and ion homeostasis. However, these compensatory mechanisms are insufficient to counteract the severe damage to photosynthetic assimilation caused by salt stress. Consequently, carbon allocation efficiency is significantly reduced, leading to a significant decrease in dry weight accumulation, as documented in previous studies [32,33]. Although the A-COS treatment failed to fully restore dry weight to the level of the unstressed control, it significantly increased dry matter accumulation to values comparable to those of the positive group treated by SA. This is consistent with the available evidence that COS can mitigate salt stress by regulating redox balance and nutrient uptake efficiency, thus supporting the maintenance of biomass under unfavorable conditions [34]. The beneficial effects may be attributed to the COS-mediated mitigation of salt-induced osmotic stress and membrane electrolyte leakage, coupled with improved cellular expansion and photosynthetic efficiency, which contributes to the increase in the fresh mass and dry mass of the plant [35,36].

The application of COS can effectively alleviate the damage caused by salt stress to plants. In this study, three distinct COSs exhibited varying efficacy in stress alleviation. Among them, the proline content in the wheat seedlings of the A-COS group increased several times, accompanied by a significant reduction in MDA levels. Previous studies have shown that COS triggers the expression of proline biosynthesis genes (i.e., OsP5CS,

OsP5CR, Osδ-OAT) and inhibits the expression of proline-degrading enzyme OsProDH [37]. We also observed an increase in proline in wheat seedlings in the A-COS group, which is supposed to be a result of upregulation of proline synthesis-related genes. Functioning as both osmolyte and antioxidant, the accumulated proline directly participates in ROS scavenging, thereby mitigating oxidative damage and reducing cellular MDA accumulation [38]. Secondly, COS treatment was found to enhance cytoplasmic free Ca^{2+} levels and activate MAPK cascades, subsequently upregulating genes associated with phytohormone signal transduction [39,40]. This hormonal signaling network activates comprehensive antioxidant defense mechanisms, effectively controlling ROS levels and attenuating membrane lipid peroxidation. In this study, it is hypothesized that the decrease in antioxidant enzyme activity in wheat seedlings after spraying the COS is possibly caused by the reduction in ROS generation, resulting in a decrease in the stress demand of antioxidant enzymes (SOD, POD, and CAT), rather than the impairment of their functions. This possibility is rooted in previous studies reporting that the activity of antioxidant enzymes in wheat seedlings is balanced with the degree of lipid peroxidation [30,41]. When ROS in plant cells are effectively controlled, the activity of antioxidant enzymes will return to the normal level, which is similar to the situation where COS derivatives of γ-aminobutyric acid ameliorate salt stress in wheat [42]. N-COS and C-COS belong to a single deacetylated COS, and structurally, they are more similar to fully acetylated CHOS. Therefore, they have a weaker stimulating effect on proline-related metabolic enzymes.

Salt stress leads to an increase in Na^+ absorption and a decrease in K^+ absorption in plants. All three COSs effectively ameliorate cellular Na^+/K^+ homeostasis in wheat seedlings, with A-COS demonstrating superior activity. Research indicates that COS enhances the expression of SOS1 and NHX2 genes in plants, both of which are critical for sodium ion regulation. SOS1 mediates cellular Na^+ efflux to the extracellular space, while NHX2 compartmentalizes intracellular Na^+ into vacuolar compartments. This dual mechanism not only mitigates sodium toxicity but also synergistically enhances K^+ uptake, collectively improving plant ionic homeostasis under saline stress conditions [43]. Notably, A-COS treatment in this case markedly enhanced K^+ assimilation and ameliorated cellular Na^+/K^+ homeostasis. The improvements mediated by COS may result from upregulation of the expression of SOS1 and NHX2 genes, encoding critical sodium compartmentalization transporters involved in the vacuoles and thereby mitigating sodium toxicity, which is consistent with findings reported in prior research [44]. The pronounced efficacy of A-COS also correlates specifically with a significant accumulation of the osmoregulatory metabolite proline in treated seedlings. This compatible solute facilitates ionic balance through the dual mechanisms of membrane stabilization and reactive oxygen species scavenging [45,46].

By analyzing the various indicators of wheat seedlings, it is found that A-COS had a more excellent induction of salt stress resistance than N-COS and C-COS, which is attributed to the change in the arrangement pattern of the COS sequence. When the acetyl groups within the COS sugar chain are deacetylated, it causes a change in the overall hydrophobicity of the sugar chain, thereby affecting the interaction between COS and the binding proteins on the plant cell surface [47]. However, the acetyl groups at the reducing ends of these three COSs in this case remain unchanged. Therefore, the relationship between the reducing-end residue of COSs and their ability to induce salt resistance activity in plants requires further investigation.

5. Conclusions

In this study, three COSs with different sequence arrangements, namely N-COS, C-COS, and A-COS, were successfully produced by chitin deacetylases NodB, *Vc*COD, and *Ar*CE4A. Their roles in alleviating salt stress in wheat seedlings were systematically investigated. The results revealed that all three COSs mitigated salt-induced damage to varying extents, but their mechanisms of action differed significantly. The multi-site deacetylated COS (A-COS) exhibited the most pronounced salt stress alleviating activity. Specifically, A-COS was used to reduce ROS in wheat seedlings by promoting the accumulation of proline and attenuating lipid peroxidation and oxidative stress. Concurrently, it promoted K^+ uptake, thereby alleviating osmotic imbalance and ion toxicity. A-COS effectively attenuated the accumulation of reactive oxygen species (ROS) in wheat seedlings through two mechanisms: (1) the enhancement of proline biosynthesis to counteract oxidative stress; (2) the suppression of lipid peroxidation cascades. At the same time, this treatment improved K^+ intake, thereby restoring ionic homeostasis by mitigating Na^+-induced osmotic imbalance and cellular toxicity. In contrast, single-site deacetylated COSs (N-COS and C-COS) did not promote proline accumulation in plants while reducing lipid peroxidation damage. Therefore, the removal of internal acetyl groups in COS chains is supposed to be critical for the induction of salt tolerance in plants, clearly establishing the positional specificity of deacetylation sites as a determinant of COS bioactivity. Unfortunately, we did not prepare COSs with deacetylation at the reducing end. COS-induced salt resistance in plants is an extremely complex process, and the exact mechanism needs to be further investigated. The effect of A-COS on plants under long-term salt exposure also needs to be conducted in order to reveal the ecological relevance of the COS application in agriculture.

Author Contributions: Conceptualization, J.L. and K.L.; methodology, J.L.; software, J.L.; validation, J.L., A.L. and Y.L.; formal analysis, J.L.; investigation, J.L.; resources, K.L.; data curation, J.L. and K.L.; writing—original draft preparation, J.L.; writing—review and editing, K.L.; visualization, K.L.; supervision, S.Z., L.S. and R.X.; project administration, S.L.; funding acquisition, K.L. All authors have read and agreed to the published version of the manuscript.

Funding: This research was supported by the National Natural Science Foundation of China (42276097), Natural Science Foundation of Shandong Province (ZR2022YQ37), Shandong Key R & D Plan, Major Scientific and Technological Innovation Project (2022CXGC020413), the Taishan Industrial Experts Program, Central guidance for local scientific and technological development funds (2024ZY0039) and the Earmarked Fund for Modern Agro-industry Technology Research System (CARS-49).

Institutional Review Board Statement: Not applicable.

Data Availability Statement: Data are contained within the article.

Conflicts of Interest: The authors declare no conflicts of interest.

Abbreviations

The following abbreviations are used in this manuscript:

COS	Chitooligosaccharides
ROS	Reactive oxygen species
SA	Salicylic acid
MDA	Malondialdehyde
PAMP	Pathogen-associated molecular pattern
A	N-acetylglucosamine
D	Glucosamine
DA	Degree of acetylation
CHOS	Chitin oligosaccharide
POD	Peroxidase
CAT	Catalase
SOD	Superoxide Dismutase

References

1. Hassani, A.; Azapagic, A.; Shokri, N. Global predictions of primary soil salinization under changing climate in the 21st century. *Nat. Commun.* **2021**, *12*, 6663. [CrossRef]
2. Fan, Y.N.; Zhang, Y.X.; Wan, M.X.; Hu, W.Y.; Chen, Z.K.; Huang, B. Plastic shed production intensified secondary soil salinization in perennial fruit production systems. *Agric. Ecosyst. Environ.* **2021**, *316*, 107469. [CrossRef]
3. Fang, X.; Mo, J.J.; Zhou, H.K.; Shen, X.F.; Xie, Y.L.; Xu, J.H.; Yang, S. Comparative transcriptome analysis of gene responses of salt-tolerant and salt-sensitive rice cultivars to salt stress. *Sci. Rep.* **2023**, *13*, 19065. [CrossRef] [PubMed]
4. Ondrasek, G.; Rengel, Z. Environmental salinization processes: Detection, implications & solutions. *Sci. Total Environ.* **2021**, *754*, 142432.
5. Liu, A.K.; Wang, M.Y.; Dong, J.J.; Yan, Z.Y.; Wang, X.; Li, J.; Song, H.F. Foliar application of exogenous salicylic acid mitigates the detrimental effects caused by salt stress in sunflower seedlings. *Ind. Crop Prod.* **2024**, *222*, 119854. [CrossRef]
6. Yang, Y.; Yao, Y.D.; Li, J.; Zhang, J.; Zhang, X.D.; Hu, L.X.; Ding, D.X.; Bakpa, E.P.; Xie, J.M. Trehalose Alleviated Salt Stress in Tomato by Regulating ROS Metabolism, Photosynthesis, Osmolyte Synthesis, and Trehalose Metabolic Pathways. *Front. Plant Sci.* **2022**, *13*, 772948. [CrossRef] [PubMed]
7. Gong, B.Q.; Wang, F.Z.; Li, J.F. Hide-and-Seek: Chitin-Triggered Plant Immunity and Fungal Counterstrategies. *Trends Plant Sci.* **2020**, *25*, 805–816. [CrossRef]
8. Wang, X.X.; Zong, Y.Y.; Zhu, Y.T.; Zhang, X.M.; Zhang, F.; Gong, D.; Prusky, D.; Bi, Y. Chitooligosaccharide inhibits growth, pathogenicity and patulin production in Penicillium expansum by downregulating global transcription factors. *Postharvest Biol. Technol.* **2025**, *222*, 113407. [CrossRef]
9. Zhao, X.L.; Wang, Y.C.; Wang, C.Y.; Zhou, W.; Ouyang, D.; Gao, S.S.; Tan, X.F.; Huang, R.; Guo, Y. Chitooligosaccharide application enhanced the growth and phytoremediation efficiency of industrial hemp in Cd-contaminated soils. *Bioresour. Technol.* **2025**, *418*, 131998. [CrossRef]
10. Zhang, X.Q.; Li, K.C.; Liu, S.; Zou, P.; Xing, R.E.; Yu, H.H.; Chen, X.L.; Qin, Y.K.; Li, P.C. Relationship between the degree of polymerization of chitooligomers and their activity affecting the growth of wheat seedlings under salt stress. *J. Agric. Food Chem.* **2017**, *65*, 501–509. [CrossRef]
11. Yu, M.L.; Wu, Q.; Zheng, D.F.; Feng, N.J.; Liang, X.L.; Liu, M.L.; Li, Y.; Mou, B.M. Plant Growth Regulators Enhance Saline–Alkali Tolerance by Upregulating the Levels of Antioxidants and Osmolytes in Soybean Seedlings. *J. Plant Growth Regul.* **2022**, *41*, 3218–3232. [CrossRef]
12. Li, K.C.; Zhang, X.Q.; Yu, Y.; Xing, R.E.; Liu, S.; Li, P.C. Effect of chitin and chitosan hexamers on growth and photosynthetic characteristics of wheat seedlings. *Photosynthetica* **2020**, *58*, 819–826. [CrossRef]
13. Fu, X.D.; Zhu, L.; Li, L.; Zhang, T.; Li, M.; Mou, H.J. Eco-friendly preparation of chitooligosaccharides with different degrees of deacetylation from shrimp shell waste and their effects on the germination of wheat seeds. *Mar. Life Sci. Technol.* **2019**, *1*, 95–103. [CrossRef]
14. Yang, G.; Hou, X.Y.; Lu, J.; Wang, M.B.; Wang, Y.H.; Huang, Y.C.; Liu, Q.T.; Liu, S.; Fang, Y.W.; Fang, Y. Enzymatic modification of native chitin and chitin oligosaccharides by an alkaline chitin deacetylase from Microbacterium esteraromaticum MCDA02. *Int. J. Biol. Macromol.* **2022**, *203*, 671–678. [CrossRef] [PubMed]
15. Linhorst, M.; Wattjes, J.; Moerschbacher, B.M. Chitin deacetylase as a biocatalyst for the selective N-acylation of chitosan oligo-and polymers. *ACS Catal.* **2021**, *11*, 14456–14466. [CrossRef]
16. Zhu, X.Y.; Zhao, Y.; Zhang, H.D.; Wang, W.X.; Cong, H.H.; Yin, H. Characterization of the specific mode of action of a chitin deacetylase and separation of the partially acetylated chitosan oligosaccharides. *Mar. Drugs* **2019**, *17*, 74. [CrossRef]
17. Hao, W.T.; Li, K.C.; Ma, Y.Z.; Li, R.F.; Xing, R.E.; Yu, H.H.; Li, P.C. Preparation and Antioxidant Activity of Chitosan Dimers with Different Sequences. *Mar. Drugs* **2021**, *19*, 366. [CrossRef]
18. Basa, S.; Nampally, M.; Honorato, T.; Das, S.N.; Podile, A.R.; Gueddari, N.E.E.; Moerschbacher, B.M. The Pattern of Acetylation Defines the Priming Activity of Chitosan Tetramers. *J. Am. Chem. Soc.* **2020**, *142*, 1975–1986. [CrossRef]
19. Gudkov, S.V.; Sarimov, R.M.; Astashev, M.E.; Pishchalnikov, R.Y.; Yanykin, D.V.; Simakin, A.V.; Shkirin, A.V.; Serov, D.A.; Konchekov, E.M.; Ogly, G.-Z.N.G.; et al. Modern physical methods and technologies in agriculture. *Phys. Usp.* **2024**, *67*, 194–210. [CrossRef]
20. Tuveng, T.R.; Rothweiler, U.; Udatha, G.; Vaaje-Kolstad, G.; Smalås, A.; Eijsink, V.G.H. Structure and function of a CE4 deacetylase isolated from a marine environment. *PLoS ONE* **2017**, *12*, e0187544. [CrossRef]
21. Bradford, M.M. A rapid and sensitive method for the quantitation of microgram quantities of protein utilizing the principle of protein-dye binding. *Anal. Biochem.* **1976**, *72*, 248–254. [CrossRef] [PubMed]

22. Kashif, M.H.; Feng, J.; Dai, R.; Fan, Y.; Xia, Y.; Liu, Z. Salicylic acid-mediated alleviation of salt stress: Insights from physiological and transcriptomic analysis in Asarum sieboldii Miq. *Chemosphere* **2024**, *362*, 142604. [CrossRef] [PubMed]
23. Bates, L.S.; Waldren, R.P.A.; Teare, I.D. Rapid determination of free proline for water-stress studies. *Plant Soil* **1973**, *39*, 205–207. [CrossRef]
24. John, M.; Röhrig, H.; Schmidt, J.; Wieneke, U.; Schell, J. Rhizobium NodB protein involved in nodulation signal synthesis is a chitooligosaccharide deacetylase. *Proc. Natl. Acad. Sci. USA* **1993**, *90*, 625–629. [CrossRef]
25. Li, X.B.; Wang, L.X.; Wang, X.S.; Roseman, S. The Chitin Catabolic Cascade in the Marine Bacterium Vibrio Cholerae: Characterization of a Unique Chitin Oligosaccharide Deacetylase. *Glycobiology* **2017**, *17*, 1377–1387. [CrossRef]
26. He, X.F.; Li, K.C.; Xing, R.E.; Liu, S.; Hu, L.F.; Li, P.C. The production of fully deacetylated chitosan by compression method. *Egypt. J. Aquat. Res.* **2016**, *42*, 75–81. [CrossRef]
27. Trombotto, S.; Ladavière, C.; Delolme, F.; Domard, A. Chemical Preparation and Structural Characterization of a Homogeneous Series of Chitin/Chitosan Oligomers. *Biomacromolecules* **2008**, *9*, 1731–1738. [CrossRef]
28. Wang, T.; Chen, Y.; Zhang, M.; Chen, J.G.; Liu, J.; Han, H.L.; Hua, X.J. Arabidopsis AMINO ACID PERMEASE1 Contributes to Salt Stress-Induced Proline Uptake from Exogenous Sources. *Front. Plant Sci.* **2017**, *8*, 2182. [CrossRef]
29. Alvan, H.A.; Jabbarzadeh, Z.; Fard, J.R.; Noruzi, P. Selenium foliar application alleviates salinity stress in sweet william (*Dianthus barbatus* L.) by enhancing growth and reducing oxidative damage. *Sci. Rep.* **2025**, *15*, 5570. [CrossRef]
30. Kul, R.; Arjumend, T.; Ekinci, M.; Yildirim, E.; Turan, M.; Argin, S. Biochar as an organic soil conditioner for mitigating salinity stress in tomato. *Soil Sci. Plant Nutr.* **2021**, *67*, 693–706. [CrossRef]
31. Shen, Z.H.; Pu, X.Z.; Wang, S.M.; Dong, X.X.; Cheng, X.X.; Cheng, M.X. Silicon improves ion homeostasis and growth of liquorice under salt stress by reducing plant Na^+ uptake. *Sci. Rep.* **2022**, *12*, 5089. [CrossRef] [PubMed]
32. Acosta-Motos, J.R.; Ortuño, M.F.; Bernal-Vicente, A.; Diaz-Vivancos, P.; Sanchez-Blanco, M.J.; Hernandez, J.A. Plant responses to salt stress: Adaptive mechanisms. *Agronomy* **2017**, *7*, 18. [CrossRef]
33. Quan, X.Y.; Liang, X.L.; Li, H.M.; Xie, C.J.; He, W.X.; Qin, Y.X. Identification and characterization of wheat germplasm for salt tolerance. *Plants* **2021**, *10*, 268. [CrossRef] [PubMed]
34. Zou, P.; Li, K.C.; Liu, S.; He, X.F.; Zhang, X.Q.; Xing, R.E.; Li, P.C. Effect of Sulfated Chitooligosaccharides on Wheat Seedlings (Triticum aestivum L.) under Salt Stress. *J. Agric. Food Chem.* **2016**, *64*, 2815–2821. [CrossRef]
35. Lamnai, K.; Anaya, F.; Fghire, R.; Zine, H.; Wahbi, S.; Loutfi, K. Impact of Exogenous Application of Salicylic Acid on Growth, Water Status And Antioxidant Enzyme Activity of Strawberry Plants (*Fragaria vesca* L.) Under Salt Stress Conditions. *Gesunde Pflanz.* **2021**, *73*, 465–478. [CrossRef]
36. Farouk, S.; Elhindi, K.M.; Alotaibi, M.A. Silicon supplementation mitigates salinity stress on Ocimum basilicum L. via improving water balance, ion homeostasis, and antioxidant defense system. *Ecotoxicol. Environ. Saf.* **2020**, *206*, 111396. [CrossRef]
37. Zhang, Y.; Fan, L.Q.; Zhao, M.Y.; Chen, Q.M.; Qin, Z.; Feng, Z.H.; Fujiwara, T.; Zhao, L.M. Chitooligosaccharide plays essential roles in regulating proline metabolism and cold stress tolerance in rice seedlings. *Acta Physiol. Plant.* **2019**, *41*, 77. [CrossRef]
38. Butt, M.; Sattar, A.; Abbas, T.; Sher, A.; Ijaz, M.; Ul-Allah, S.; Shaheen, M.R.; Kaleem, F. Foliage applied proline induces salt tolerance in chili genotypes by regulating photosynthetic attributes, ionic homeostasis, and antioxidant defense mechanisms. *Hortic. Environ. Biotechnol.* **2020**, *61*, 693–702. [CrossRef]
39. Lu, H.; Zhao, X.M.; Wang, W.X.; Yin, H.; Xu, J.G.; Bai, X.F.; Du, Y.G. Inhibition effect on tobacco mosaic virus and regulation effect on calreticulin of oligochitosan in tobacco by induced Ca^{2+} influx. *Carbohyd. Polym.* **2010**, *82*, 136–142. [CrossRef]
40. Brulé, D.; Villano, C.; Davies, L.J.; Trdá, L.; Claverie, J.; Héloir, M.C.; Chiltz, A.; Adrian, M.; Darblade, B.; Tornero, P.; et al. The grapevine (Vitis vinifera) LysM receptor kinases VvLYK1-1 and VvLYK1-2 mediate chitooligosaccharide-triggered immunity. *Plant Biotechnol. J.* **2019**, *17*, 812–825. [CrossRef]
41. Al-Huqail, A.A.; Aref, N.M.A.; Khan, F.; Sobhy, S.E.; Hafez, E.E.; Khalifa, A.M.; Saad-Allah, K.M. Azolla filiculoides extract improved salt tolerance in wheat (*Triticum aestivum* L.) is associated with prompting osmostasis, antioxidant potential and stress-interrelated genes. *Sci. Rep.* **2024**, *14*, 11100.
42. Wang, W.Y.; Liu, S.; Yan, M.Y. Synthesis of γ-Aminobutyric Acid-Modified Chitooligosaccharide Derivative and Enhancing Salt Resistance of Wheat Seedlings. *Molecules* **2022**, *27*, 3068. [CrossRef]
43. Zou, P.; Li, K.C.; Liu, S.L.; Xing, R.E.; Qin, Y.K.; Yu, H.H.; Zhou, M.M.; Li, P.P. Effect of chitooligosaccharides with different degrees of acetylation on wheat seedlings under salt stress. *Carbohyd. Polym.* **2015**, *126*, 62–69. [CrossRef] [PubMed]
44. Chatelain, P.G.; Pintado, M.E.; Vasconcelos, M.W. Evaluation of chitooligosaccharide application on mineral accumulation and plant growth in Phaseolus vulgaris. *Plant Sci.* **2014**, *215*, 134–140. [CrossRef] [PubMed]
45. Hou, Y.M.; Xu, X.S.; Kong, X.L. K^+-Selectivity Due to Coordination with a D_{4d}-Symmetric Homochiral Proline Octamer Verified by Mass Spectrometry and Infrared Photodissociation Spectroscopy. *J. Phys. Chem. Lett.* **2023**, *14*, 2660–2664. [CrossRef] [PubMed]

46. Bojórquez-Quintal, E.; Ruiz-Lau, N.; Velarde-Buendía, A.; Echevarría-Machado, I.; Pottosin, I.; Martínez-Estévez, M. Natural variation in primary root growth and K^+ retention in roots of habanero pepper (*Capsicum chinense*) under salt stress. *Funct. Plant Biol.* **2016**, *43*, 1114–1125. [CrossRef]
47. Hayafune, M.; Berisio, R.; Marchetti, R.; Silipo, A.; Kayama, M.; Desaki, Y.; Arima, S.; Squeglia, F.; Ruggiero, A.; Tokuyasu, K.; et al. Chitin-induced activation of immune signaling by the rice receptor CEBiP relies on a unique sandwich-type dimerization. *Proc. Natl. Acad. Sci. USA* **2014**, *111*, E404–E413. [CrossRef]

polymers

MDPI

Article

Development of Curcumin-Loaded TiO_2-Reinforced Chitosan Monofilaments for Biocompatible Surgical Sutures

Fatma Demirci

Department of Polymer Materials Engineering, Faculty of Engineering and Natural Sciences, Bursa Technical University, 16310 Bursa, Türkiye; fatma.demirci@btu.edu.tr

Abstract: Sutures provide mechanical support for wound closure after various traumas and surgical operations. Absorbable sutures are increasingly favored as they eliminate the need for secondary procedures and minimize additional damage to the wound site. In this study, chitosan sutures were produced using the dry jet–wet spinning method, achieving number 7-0 sutures (approximately 76 μm diameter) with a homogeneous surface. FTIR analysis demonstrated molecular interactions between chitosan and TiO_2 or curcumin, confirming successful incorporation. The addition of 3% TiO_2 increased the tensile strength of chitosan sutures by 12.32%, reaching 189.41 MPa. Morphological analysis revealed smooth surfaces free of pores and bubbles, confirming the production of high-quality sutures. Radical scavenging activity analysis showed that curcumin-loaded sutures exhibited 43% scavenging ability after 125 h, which was significantly higher compared to pure chitosan sutures. In vitro antibacterial tests demonstrated that curcumin-loaded sutures provided 98.87% bacterial inactivation against *S. aureus* within 24 h. Additionally, curcumin release analysis showed a cumulative release of 77% over 25 h. The bioactivity of the sutures was verified by hydroxyapatite layer formation after incubation in simulated body fluid, supporting their potential for tissue regeneration. These findings demonstrate that TiO_2 reinforcement and curcumin loading significantly enhance the functional properties of chitosan sutures, making them strong candidates for biocompatible and absorbable surgical applications.

Keywords: dry jet–wet spinning; chitosan; suture; drug release; antioxidant activity; bioactivity

1. Introduction

Various medical devices can be used to hold tissues together after trauma or surgery. Although significant advances have been made in tissue adhesives and other mechanical wound closure devices such as surgical staples and skin closure tapes, surgical sutures are still the most preferred method of wound closure in both human and veterinary medicine due to their easy accessibility and applicability [1,2]. Although not all of them will provide all the features, an ideal suture provides appropriate mechanical strength, causes minimal tissue damage and reaction, reduces the risk of infection, is easy to use, and is relatively inexpensive. Sutures that provide the necessary mechanical support to the wound during the healing period are preferred in many medical applications with their stated superior features and therefore the demand for sutures increases by millions of dollars every year [3]. Sutures can be classified in various categories depending on the material used (natural or synthetic), their degradability (absorbable or non-absorbable), and their physical configuration (monofilament, multifilament, twisted, or braided) [3,4].

The most basic features expected from a suture are that it reinforces the wound closure externally and is biocompatible to support wound healing [5,6]. These requirements vary

 https://doi.org/10.3390/polym17040484

depending on the type of wound and its location. For example, while tendons take weeks to heal, epithelial tissue and muscles can heal within a few days [3,7]. Although non-absorbable sutures generally have higher durability, they can cause complications such as chronic inflammation, infection, and tissue damage [6]. On the other hand, they need to be removed from the patient with a second operation. For all these reasons, the development of absorbable sutures is seen as an important development in the field of surgical sutures [2,3]. Absorbable surgical sutures can be produced from natural polymers such as catgut [8], collagen [9], and chitin [10] fibers or synthetic polymers such as polyglycolic acid [11], poly(p-dioxanone) [12], poly(lactic-co-glycolic acid) [13], poly(trimethylene carbonate) [14], and polycaprolactone [15]. Natural origin sutures stand out due to their properties such as antibacterial, antifungal, and preventing scar formation [4,16].

The limited solubility of chitin in both aqueous and organic solutions restricts its practical applications. However, chitosan, which is derived from chitin through deacetylation, offers improved solubility and is therefore more suitable for biological applications [17]. Due to its excellent biocompatibility [18], biodegradability [19], low toxicity [20], antimicrobial [21], antioxidant [22], and anticancer [23] properties, chitosan is widely regarded as a promising polymer for biomedical and pharmaceutical applications [24]. Despite these advantages, the low mechanical strength of chitosan-based sutures restricts their direct use in surgical applications. As a result, chitosan is predominantly applied as a coating material on commercial sutures. Nevertheless, recent studies have explored the direct production of sutures from chitosan, aiming to overcome its mechanical limitations while preserving its beneficial biological properties. Montenegro and Godeiro used QiGel® chitosan fibers being produced by Medovent GmbH as a suture material on rats and demonstrated the superiority of chitosan sutures by comparing their biodegradable and bacteriostatic properties with different sutures [25]. Silva et al. produced drug-loaded wet-spun chitosan sutures and investigated the morphological, biodegradation, drug release, mechanical, and cytotoxic properties of the produced sutures. It was stated that N-acetyl-D-glucosamine additive reduces mechanical performance but improves properties such as biodegradability, biocompatibility, and extended drug release of chitosan sutures, making them a good candidate for absorbable sutures [4]. Perrin et al. produced wet-spun chitosan sutures from high molecular weight fungal and shrimp chitosan and low molecular weight shrimp chitosan. They revealed the chosen chitosan and solvent effect on mechanical properties of the suture. They also examined the morphological properties and biocompatibility of the produced sutures [26]. Tan et al. prepared ultra-high molecular weight chitosan with different deacetylation degrees and molecular weights, and then produced chitosan sutures with a wet-spinning process. The effects of the deacetylation degree and molecular weight of chitosan on the morphological, mechanical, swelling degree, enzymatic degradation, and cytotoxic properties of the produced sutures were investigated [3].

As mentioned, although chitosan exhibits very good properties in terms of biomedical applications, the most important parameter limiting the industrialization of chitosan in suture applications is its low mechanical performance. To address this limitation, various additives such as nanoclay [27], carbon nanotubes [28], or inorganic nanoparticles [29] have been explored in the literature to enhance the mechanical properties of chitosan. Among these, TiO_2 was selected as the sole additive in this study due to its well-documented ability to significantly improve both the mechanical strength and bioactivity of polymeric materials. TiO_2 nanoparticles form covalent bonds with chitosan, leading to enhanced tensile properties without compromising biocompatibility. Additionally, TiO_2 has demonstrated superior biocompatibility [30], bioactivity [31], and antimicrobial [30,31] potential in biomedical applications compared to other additives. While carbon nanotubes may pose cytotoxicity risks, nanoclay is primarily used as a carrier for bioactive agents rather than directly enhancing bioactivity. In contrast, TiO_2 not only reinforces the mechanical properties of the suture but also actively contributes to its bioactive characteristics [32–36]. However, it should be noted that the effects of TiO_2 can vary depending on factors such as particle size, concentration, and surface chemistry. In some cases, excessive TiO_2 content has been reported to cause aggregation, leading to a reduction in mechanical performance, or to interfere with cellular interactions, affecting bioactivity negatively [37–39].

Curcumin is a phytopolyphenol obtained from the *Curcuma longa* plant [40]. This bioactive compound is a multifunctional active agent known with antibacterial [41], antifungal [42], antiviral [43], antioxidant [44], anti-inflammatory [44], anticoagulant [45], antiatherosclerotic [46], anticarcinogenic [47], and hypoglycemic [48] effects. Due to these therapeutic effects, it is preferred for use in pharmaceutical and biomedical fields.

This study demonstrates the production of chitosan sutures using the dry jet–wet spinning method, which has not been previously reported in the literature for chitosan-based surgical sutures. TiO_2 was incorporated to enhance the mechanical strength of chitosan, addressing its limitations in suture applications. Once the desired mechanical properties were achieved, curcumin was added to improve the therapeutic potential of the sutures. The obtained sutures were characterized in terms of their morphological, physicochemical, and mechanical properties. Additionally, the effects of TiO_2 and curcumin on radical scavenging activity, biocidal performance, and bioactivity were evaluated. The release profile of curcumin from the sutures was also examined to assess its potential for sustained therapeutic effects.

2. Materials and Methods

2.1. Materials

Chitosan (85% deacetylated) was supplied from Alfa Aesar (Karlshure, Germany) and acetic acid (99.8% purity) was purchased from Sigma-Aldrich (Steinheim, Germany). To prepare the coagulation bath, sodium hydroxide (NaOH) was acquired from Sigma-Aldrich (Prague, Czech Republic) and ethanol was obtained from Alkomed (Kocaeli, Türkiye). TiO_2 in rutile phase used as additive material was synthesized by Duman and her team at Bursa Technical University Metallurgical Materials Engineering Laboratory. The TiO_2 has 254.2 nm particle size, 18.9 m^2/g surface area, and 4.22 g/cm^3 density [49]. The other additive curcumin was obtained from Gemma (Kocaeli, Türkiye) in powder form. For radical scavenging activity tests, 2,2-Diphenyl-1-(2,4,6-trinitrophenyl)hydrazyl (DPPH) was purchased from Sigma-Aldrich (Steinheim, Germany) and methanol was obtained from Merck (Darmstadt, Germany). Reagents to prepare the simulated body fluid (SBF) solution, sodium chloride (NaCl), potassium chloride (KCl), calcium chloride dihydrate ($CaCl_2{\cdot}2H_2O$), magnesium chloride hexahydrate ($MgCl_2{\cdot}6H_2O$), and sodium dihydrogen phosphate monohydrate ($NaH_2PO_4{\cdot}H_2O$), were purchased from Merck (Darmstadt, Germany) and sodium bicarbonate ($NaHCO_3$) was obtained from Sigma-Aldrich (Steinheim, Germany). Phosphate-buffered saline (PBS, pH 7.4) was obtained from Gündüz Kimya™ (İstanbul, Türkiye). Tween 80 was acquired from KimyaLab (İstanbul, Türkiye).

2.2. Production of the Sutures

The dry jet–wet spinning process was used to produce chitosan sutures. Chitosan was dissolved in an acetic acid solution, taking into account both the required acid concentration and the degree of deacetylation, as referenced in Perrin et al. [26]. Based on literature studies, a 4% chitosan solution was selected for fiber spinning and dissolved in 0.239 mol/L acetic acid by stirring overnight at room temperature [4,26,50–52]. TiO_2 was then added at concentrations of 1%, 3%, and 5% by weight of chitosan, and the solution was sonicated for 15 min to ensure homogeneous dispersion before being mixed overnight. For the production of curcumin-loaded sutures, 1% curcumin (by weight of chitosan) was added to the solution and mixed for 45 min at room temperature. Throughout the preparation of the suture spinning solutions, the total solids concentration was maintained at 4 wt%. The compositions of the spinning solutions of the produced sutures are given in Table 1.

Table 1. Composition of the spinning solutions of the produced sutures.

Sample Code	Chitosan (g)	TiO_2 (g)	Curcumin (g)
CS	0.800	0	0
CS/1T	0.792	0.008	0
CS/3T	0.776	0.024	0
CS/5T	0.760	0.040	0
CS/3T/C	0.768	0.024	0.008

Sutures were produced using a custom-made spinning unit. The prepared solutions were transferred to a 20 mL syringe and pumped into the coagulation bath at pH 13 (1:1 mL/mL; ethanol and 10% aqueous solution of NaOH) with a flowing rate of the solution of 2.5 mL/min using a syringe pump (IPS 12, Inovenso (İstanbul, Türkiye)). The fibers entered the coagulation bath after passing a 105 mm air gap, and passed through the washing bath by the help of a 61 mm diameter perforated drum. Then, the fibers were wrapped around a 620 mm diameter wheel and kept in the washing bath for approximately 1 h until the pH reached close to 7. Subsequently, the fibers were dried overnight at room temperature by stretching. While the chitosan suture produced without additives was called CS, the TiO_2 added sutures were named CS/1T, CS/3T, and CS/5T depending on the additive ratio they contained. The suture produced by adding curcumin was named CS/3T/C.

2.3. Characterizaton of the Produced Sutures

The structure of the produced sutures and the effects of the TiO_2 and curcumin addition were observed with Fourier-transform infrared spectroscopy (FTIR). FTIR spectra of the produced sutures were recorded by a PerkinElmer Spectrum Two spectrometer (Shelton, CT, USA) in the scanning range of 400 to 4000 cm^{-1} and 16 scans were made at 4 cm^{-1}.

The surface morphology of the produced sutures was observed by field emission scanning electron microscopy (FE-SEM, Carl-Zeiss/Gemini 300 (Oberkochen, Germany)) before and after incubation in SBF. Ten nm gold palladium was coated on the surface of the suture samples and coated samples were imaged with 200× and 500× magnifications.

Mechanical properties of the produced sutures were determined by uniaxial tensile tests according to ASTM D2256 standard [53] with 2 mm/min strain rate, similar to previous research [4].

In vitro swelling behavior of the produced sutures was determined by the gravimetric method. The dry weights of the samples cut from the produced sutures were weighed (W_D) and immersed in 5 mL of phosphate-buffered saline (PBS, pH 7.4, 37 °C) solution. After a 24 h incubation period, the samples were taken from solution and the wet weights of the samples (W_W) were determined after the fluid on the sample surface was gently blotted. The swelling degrees (W_S) of the samples for different incubation times were calculated using Equation (1):

$$(W_S) = (W_W - W_D), \tag{1}$$

Radical scavenging activity (RSA) of the produced sutures was observed by the DPPH method. A 1 mM DPPH solution was prepared by dissolving 3.94 mg DPPH in 100 mL methanol. An amount of 20 mg of the produced sutures was placed in dark test tubes containing 4 mL of methanol and 0.5 mL of prepared methanolic DPPH solution was added to each tube and the samples were stored in a darkroom at room temperature. The absorbance of reaction solution was measured daily with a UV–Visible spectrophotometer (Scinco-NEOSYS200 (Taipei, Taiwan)) at 517 nm and the RSA% of samples was calculated using Equation (2):

$$RSA\ (\%) = ((A_{DPPH} - A_{SAMPLE})/A_{DPPH}) \times 100, \quad (2)$$

where A_{DPPH} and A_{SAMPLE} were the absorbance of the blank and reaction solutions, respectively.

To determine curcumin loading efficiency, a known amount of CS/3T/C suture sample and curcumin was dissolved in a 10.5 mL (10:10:1) ethanol/distilled water/acetic acid solution. The absorbance of the obtained solutions was measured with a UV–Visible spectrophotometer (Scinco-NEOSYS200) at 431 nm and the loading efficiency (LE%) was calculated using Equation (3):

$$LE\ (\%) = ((m_C \times A_{CS/3T/C})/(m_{CS/3T/C} \times A_C)) \times 100, \quad (3)$$

where m_C and $m_{CS/3T/C}$ were the weight of the curcumin and curcumin-loaded suture sample dissolved in ethanol/distilled water/acetic acid solution, and A_C and $A_{CS/3T/C}$ were the absorbance of the curcumin and curcumin-loaded suture sample solutions, respectively.

For curcumin release analysis, firstly the amount of curcumin in 10 mg suture was calculated according to the value obtained from the LE result and accordingly the calibration curve was obtained by determining the total curcumin ratio in the suture. In order to determine the curcumin release, 10 mg of CS/3T/C sample was immersed in 10 mL of PBS solution containing 5% Tween 80 for 48 h at 37 °C. Since curcumin was insoluble in pure PBS solution, Tween 80 was added to the solution to dissolve free-released curcumin. Samples were taken in solution at different times and measurements were taken with a UV–Visible spectrophotometer at 431 nm. The experiment was performed in triplicate.

The in vitro bioactivity of the produced sutures (CS, CS/3T, CS/3T/C) was determined by immersion in 10 × SBF (adjusted at pH 7.4 with $NaHCO_3$) which contains 10 times the calcium and phosphate ion concentrations of human plasma and being prepared according to a previous study [54]. After 3 and 14 days of incubation in 10 × SBF, the suture samples were removed from the solution, gently washed with deionized water, and dried at 37 °C. After the incubation period, the surface morphology of the sutures was observed by SEM. The Ca/P ratio of the mineralized layer of the CS/3T/C sample was analyzed using an Energy Dispersive X-ray Spectrometer (EDS, Bruker™ XFlash 6I100 (Berlin, Germany)). The chemical composition of the HA layer on the CS/3T/C suture was determined by FTIR analysis.

Biocidal efficacy performance of the produced sutures was determined under dynamic contact conditions according to ASTM E2149. The suture samples were tested against *Staphylococcus aureus* (ATCC 6538) and *Escherichia coli* (ATCC 35218). Briefly, for antibacterial activity evaluation, 2.5 mg suture samples were placed in 2 mL known concentration bacterial suspension (approximately 5×10^5 CFU/mL) and shaken. For 3, 6, and 24 h contact times, 100 µL of bacterial suspensions were taken and diluted with phosphate-buffered solution. The diluted solutions were spread onto Mueller-Hilton II agar media and incubated for 24 h at 37 °C. At the end of the incubation period, viable bacterial colonies were counted and the inhibition percentage (%) was calculated using Equation (4):

$$\text{Reduction rate of bacteria}\ (\%) = ((A - B)/A) \times 100, \quad (4)$$

where A is the number of bacteria recovered from the test sample at "0" contact time, and B is the number of bacteria recovered from test samples incubated for specific periods of time, respectively.

3. Results and Discussion

3.1. Influence of the TiO_2 Addition on the Mechanical Properties of the Sutures

The mechanical performance of the suture is of great importance both in terms of ease of operation and keeping the wound closed until tissue regeneration after application. Despite its advantages such as biocompatibility, non-toxicity, and biodegradability, chitosan is studied in the literature as a coating on sutures rather than direct suture applications due to its low mechanical properties [1]. In recent years, studies have been carried out on the production of chitosan sutures with improved mechanical strength and functional properties [4,26]. As known in the literature, TiO_2 nanoparticles transform into titanium cation (Ti^{4+}) in an acidic environment and form covalent bonds with –OH and –NH_2 groups of the chitosan [55]. In the first part of this study, TiO_2 was added to the chitosan suture in order to increase its mechanical properties. TiO_2 addition effects on the mechanical properties of the sutures are given in Table 2. The tensile strength of the CS suture was 168.63 MPa and the addition of 1 wt% and 3 wt% TiO_2 nanoparticles to the suture structure caused an increase in both tensile strength and elongation at break. This increase could be related to the molecular interaction between the TiO_2 nanoparticles and chitosan [55]. As seen in Table 2, the tensile strength of the CS/3T suture reached 189.41 MPa, demonstrating a 12.32% improvement compared to the unmodified CS suture. However, when the TiO_2 content increased to 5 wt%, the tensile strength dropped to 148.83 MPa, which is lower than the CS suture. Similar trends have been observed in previous studies, where excessive TiO_2 addition to chitosan-based films led to aggregation, reducing mechanical performance [38,39].

Table 2. Mechanical properties of the produced sutures.

Sample	Breaking Strength (N)	Tensile Strength (MPa)	Elongation at Break (%)
CS	0.74 ± 0.02	168.63 ± 5.17	2.55 ± 0.18
CS/1T	0.77 ± 0.05	176.17 ± 12.23	2.63 ± 0.52
CS/3T	0.84 ± 0.04	189.41 ± 8.21	3.37 ± 0.49
CS/5T	0.66 ± 0.02	148.83 ± 5.39	2.77 ± 0.30
CS/3T/C	0.57 ± 0.01	128.10 ± 2.67	2.5 ± 0.01

According to the United States Pharmacopeia (USP) standards, the minimum breaking force required for 7-0 surgical sutures is 0.785 N. The breaking strength of the CS suture was measured as 0.74 N, which is slightly below the USP requirement. However, with the addition of 1 wt% and 3 wt% TiO_2, the breaking strength increased to 0.77 N and 0.84 N, respectively, surpassing the USP standard. This indicates that the CS/3T suture meets the mechanical strength criteria defined for 7-0 surgical sutures, making it a viable candidate for clinical applications. Conversely, the CS/5T suture exhibited a breaking strength of 0.66 N, which is lower than the USP requirement. This decrease in mechanical strength can be attributed to TiO_2 nanoparticle agglomeration, which disrupted the polymer network and reduced overall tensile performance.

In addition to tensile strength, elongation at break plays a critical role in the mechanical behavior and functional performance of surgical sutures. Sutures need to retain a degree of flexibility to accommodate tissue swelling and contraction during the healing process. When swelling occurs, the suture should stretch to prevent excessive tension on the wound and when the swelling subsides, the suture should return to its original shape and length to maintain tissue approximation [4]. The CS/3T suture had the highest elongation at break (3.37%), indicating its improved flexibility compared to the CS (2.55%) and CS/5T (2.77%) samples. This suggests that the CS/3T suture not only meets mechanical strength requirements, but also retains an appropriate level of elasticity, making it a suitable candidate for clinical applications requiring sutures with improved adaptability to dynamic wound environments. Furthermore, these findings emphasize that chitosan-based sutures reinforced with an optimized concentration of TiO_2 can achieve the necessary mechanical performance for surgical applications while preserving their biocompatibility and biodegradability.

Since the tensile strength of the chitosan suture was increased by approximately 12 wt% with 3 wt% addition of TiO_2, curcumin was added to the CS/3T sample in the continuation of the study, and the following analyses were carried out on the CS, CS/3T, and CS/3T/C samples.

3.2. Morphological Properties of the Produced Sutures

The surface morphology of the produced sutures was analyzed using SEM. Figure 1 shows that both the blank and doped sutures exhibited a smooth and homogeneous surface morphology without bubbles or pores. The cross-sections of the sutures were round-shaped, indicating uniformity in fiber formation. The measured diameters of the sutures were 76.69 μm for CS, 78.15 μm for CS/3T, and 77.10 μm for CS/3T/Cur, all of which fall within the 7-0 suture range (70–99 μm) as defined by the USP. The SEM images confirm that TiO_2 nanoparticles and curcumin were successfully incorporated into the chitosan suture structure, maintaining a uniform distribution.

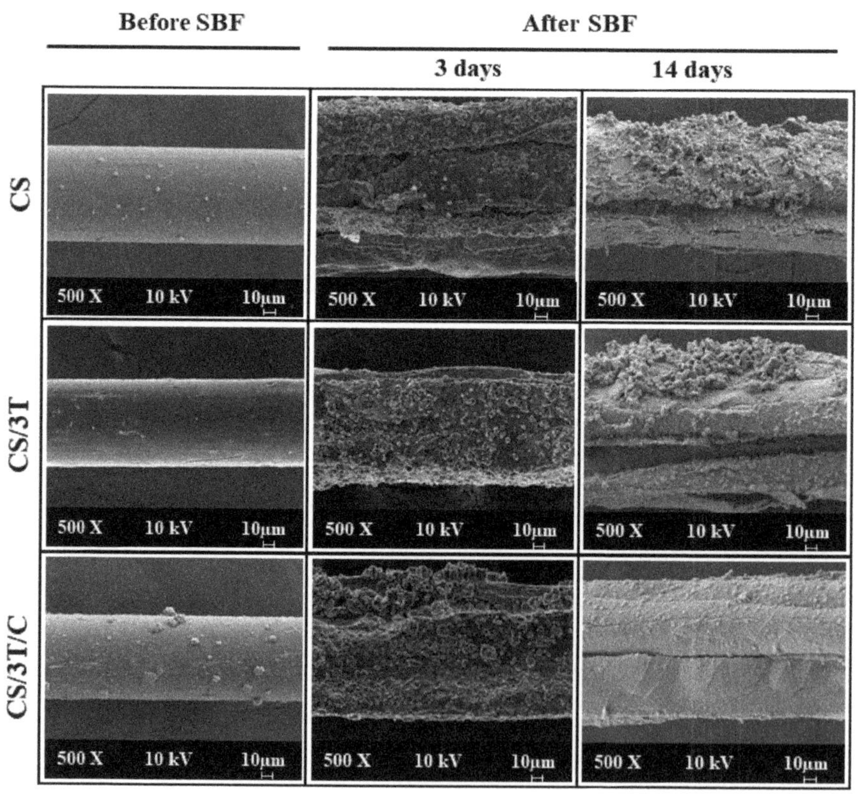

Figure 1. SEM images of the produced sutures before and after SBF (3 days and 14 days incubation).

In the studies in the literature, working at a high chitosan concentration such as 4 wt% and using ethanol in the coagulation bath leads to rapid and uncontrolled coagulation, often resulting in surface irregularities [26,51,56]. Additionally, coagulation in alkaline bath conditions further accelerates the phase separation process by deprotonating the charged amino groups of chitosan, allowing acetic acid and ethanol to demix more easily [26,57,58]. This rapid coagulation disrupts polymer chain organization, leading to uneven fiber morphology [26,51,56]. However, as demonstrated in Figure 1, SEM images confirm that the fiber surfaces obtained in this study are smooth and free of irregularities, despite using a high chitosan concentration (4%) and NaOH in the coagulation bath. This is likely due to the dry jet–wet spinning process, which introduces a critical air-gap stage before coagulation. Unlike wet spinning, where the polymer solution is extruded directly into the bath, the 105 mm air gap in this study allows polymer chains to partially align and stabilize under the influence of gravity before coagulation occurs. This controlled pre-orientation minimizes abrupt phase separation and prevents the coagulant from deeply penetrating the polymer matrix, leading to a more uniform and smoother fiber surface.

3.3. Physicochemical and Mechanical Properties of the Produced Sutures

The FTIR spectra of neat, TiO_2, and curcumin-added sutures are given in Figure 2. The strong and broad band in the region 3471–3244 cm^{-1} corresponds to symmetric vibration of N-H and O-H. The absorption bands at 2931 and 2860 cm^{-1} can be attributed to methylene and methyl groups C-H stretching vibrations. The bands at 1650 cm^{-1} were characteristic of the C=O stretching of amide I, proving the presence of remaining N-acetyl groups in the structure of chitosan. The bands at 1585 cm^{-1} and 1156 cm^{-1} were corresponding to N-H bending of amide II and asymmetric stretching of the C-O-C bridge in the glucosamine ring, respectively. These bands overlap with the spectra of chitosan samples reported in previous studies in the literature [59,60]. It is seen that the intensities of the N-H, O-H band vibrations in the 3471–3244 cm^{-1} region and amide II band vibrations at 1585 cm^{-1} decrease with the addition of TiO_2 and curcumin to the chitosan suture structure. These decreases were caused by the interaction between the additives and chitosan [61]. The characteristic Ti–O stretching vibration of TiO_2 typically appears around 600 cm^{-1} in FTIR spectra [62]. However, in the CS/3T suture, this band is observed at 520 cm^{-1}, indicating a slight shift. This shift suggests an interaction between TiO_2 and the chitosan matrix, likely due to covalent or hydrogen bonding. The presence of this band confirms the successful incorporation of TiO_2 within the suture structure. Additionally, a new vibrational band at 1502 cm^{-1} was observed exclusively in the CS/3T/C sample, which was not present in the other sutures. This band corresponds to the C=O stretching vibration of the benzene ring in the curcumin structure, confirming the successful incorporation of curcumin within the suture matrix [63].

The mechanical properties of the produced sutures are given in Table 2. The tensile strength of the CS suture was 168.63 MPa, while it was 150.90 MPa in the literature for wet-spun fibers produced at the same chitosan concentration [51]. This difference in mechanical performance is due to the production carried out with the dry jet–wet spinning process in this study. Similar results were seen in the study of Notin et al., where it was stated that physical treatment during the dry jet process improved the mechanical performance without crosslinking or post-treatment [64]. As mentioned in Section 3.1, the mechanical performance of the suture increased with the covalent bonds established between chitosan and TiO_2 with the TiO_2 addition. However, as shown in Table 2, the addition of curcumin led to a significant decrease in mechanical performance, with the tensile strength dropping to 128.10 MPa in the CS/3T/C sample. While the specific additive differs, a similar trend was observed in the study of Silva et al. It was observed that the tensile strength of N-Acetyl-D-Glucosamine-loaded chitosan fibers decreased by approximately 30% [4].

This decrease in mechanical performance was attributed to the decrease in secondary bond interactions caused by additive–polymer interactions [4]. This comparison is not intended to equate the two materials directly but rather to illustrate that the introduction of certain bioactive additives into a polymer matrix can disrupt intermolecular interactions, leading to reduced mechanical performance. Likewise, in this study, the weakening effect can be attributed to curcumin-induced disruptions in the polymer network, leading to reduced polymer chain interactions. Additionally, the increased swelling behavior further contributed to the mechanical deterioration. As previously mentioned, according to the USP standards, the minimum breaking force required for 7-0 surgical sutures is 0.785 N. The CS/3T/C suture exhibited the lowest breaking strength at 0.57 N, significantly below this threshold. This reduction in breaking strength correlates with the decrease in tensile strength and suggests that the mechanical stability of the CS/3T/C suture was compromised due to curcumin incorporation.

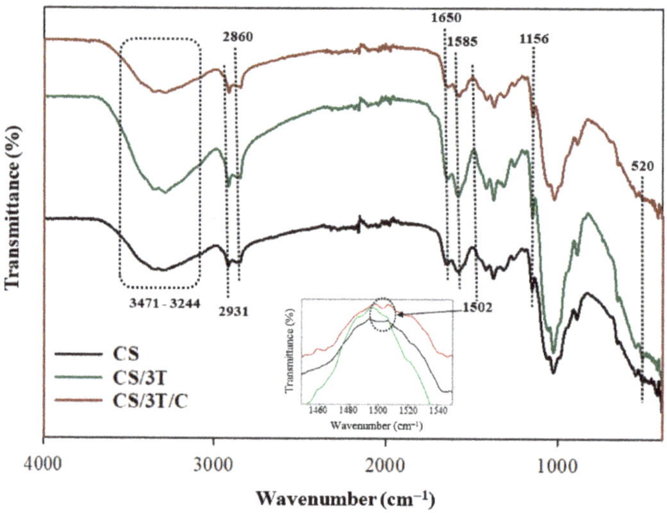

Figure 2. FTIR spectra of the produced suture samples.

Elongation at break is another critical factor that influences the functionality of surgical sutures. Sutures must exhibit sufficient flexibility to accommodate tissue movements, particularly during swelling and contraction in the wound area. The CS/3T/C suture exhibited an elongation at break of 2.5%, indicating a slight reduction in flexibility compared to the CS suture. This decrease in elongation suggests that curcumin incorporation may have altered the polymer network, potentially reducing its ability to accommodate tissue dynamics.

On the other hand, the degree of swelling significantly affects the mechanical integrity of the suture. The swelling rates of the produced sutures after 24 h of PBS incubation were determined as 124% for CS, 99% for CS/3T, and 180% for CS/3T/C. As shown in Table 2, the addition of curcumin resulted in a decrease in tensile strength, which correlates with the significant increase in swelling behavior. In the study of Albanna et al., it was stated that increased swelling in chitosan-based sutures leads to a reduction in mechanical performance due to water absorption, which disrupts polymer chain interactions and weakens structural integrity [65]. Similarly, studies have demonstrated that curcumin incorporation into polymeric matrices can enhance swelling behavior due to its interaction with the polymer network. For instance, in the study of Ciftci et al., PVA/chitosan composite mats loaded with curcumin showed that curcumin increased the swelling capacity by affecting the polymer network structure and hydrophilicity, which in turn influenced mechanical stability [66]. Furthermore, in the study of Handy and Saeed, it was found that curcumin incorporation into chitosan-based nanocomposites altered the hydration properties of the polymer, resulting in increased swelling ratios and decreased mechanical strength. This effect was attributed to curcumin-induced structural modifications within the chitosan matrix, leading to reduced polymer chain interactions and higher water uptake [67]. Similarly, in this study, the highest swelling ratio (180%) was observed in curcumin-loaded sutures (CS/3T/C), which also exhibited the lowest tensile strength and breaking strength. This suggests that the excessive swelling behavior induced by curcumin further contributed to the mechanical deterioration of the sutures.

3.4. Radical Scavenging Activities of the Produced Sutures

Free radicals cause many diseases and aging by damaging cells [68]. The use of antioxidant agents helps prevent the progression of diseases by capturing free radicals in human and animal cells [69,70]. To determine the antioxidant activity of the produced sutures, DPPH was used as a stable free radical that acts as a hydrogen and electron acceptor. As seen in Figure 3, the scavenging ability of CS suture was determined to be approximately 22.36%, and this value was very close to the scavenging ability determined in the study of Shanmugam et. al., who studied the same chitosan concentration (5 mg/mL) [71]. The scavenging mechanism of chitosan was attributed to an unshared electron pair of electrons of nitrogen in the C-2 position of the chitosan. In an acidic medium, this unshared electron catches the released proton and forms NH_3^+. Hydrogen ions of NH_3^+ can stabilize the free radicals causing various diseases by reacting with them [72,73]. The DPPH scavenging activity of the CS/3T was decreased to approximately 9.95%. This decrease in the DPPH scavenging activity of CS/3T is thought to be due to the binding of some –NH2 groups at the C-2 position with TiO_2, which are responsible for the scavenging activity in chitosan. In addition to the blocking of the active groups of chitosan, the weak antioxidant activity of TiO_2, as stated in the literature, was also effective in this decrease [74,75]. The CS/3T/CUR suture shows time-dependent activity, which confirms that the radical scavenging activity of the sutures was linked to the release of the active ingredient from the structure, similar to the literature [76,77]. As seen in the figure, while the DPPH scavenging activity of the CS/3T/CUR suture was initially between those of the CS and CS/3T sutures, it caught up with the CS suture at the 66th hour with the release of curcumin in the structure. It then continued to increase until the 125th hour and stabilized at approximately 43%.

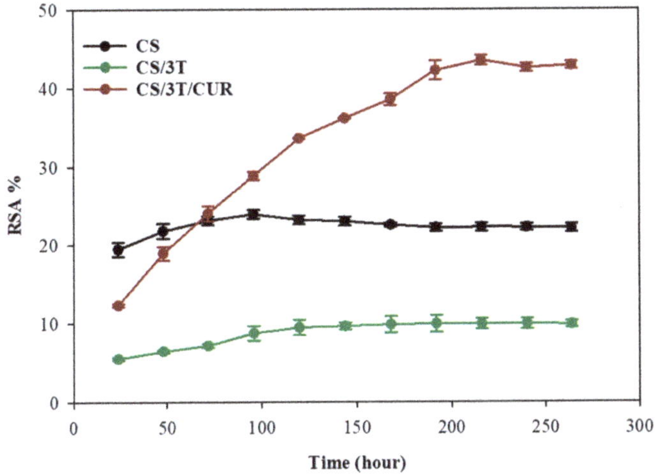

Figure 3. Radical scavenging activity of the produced sutures at different time intervals.

3.5. In Vitro Characterization of the Produced Sutures

Biocidal performance of the produced sutures against *S. aureus* was evaluated. The CS, CS/3T, and CS/3T/C suture swatches were challenged with 10^5 CFU concentration *S. aureus* and the results are given in Figure 4A. Although chitosan is known as antibacterial in the literature, the amine group must be positively charged in order for chitosan to show antibacterial activity. Amine groups can only affect negatively charged groups on the bacterial cell surface when they are positively charged [78,79]. In other words, the antibacterial properties of chitosan are directly affected by the pH of the environment, and for this reason, the antibacterial activity of the suture produced only from chitosan in this study reached only 43% after 12 h. As shown in Figure 4A, the TiO_2-doped sample showed similar antibacterial activity to CS. It is well known that TiO_2 nanoparticles exhibit antibacterial properties mainly through a photocatalytic mechanism, where they generate reactive oxygen species (ROS), including hydroxyl radicals (–OH) and superoxide anions (O_2^-), under UV or visible light irradiation [38,62,80,81]. These ROS species are highly reactive and can damage bacterial membranes, proteins, and DNA, ultimately leading to cell death. However, in the absence of light, the generation of ROS is significantly reduced, limiting the antibacterial activity of TiO_2. As the bacterial tests in this study were performed in a dark environment, TiO_2 could not be activated to produce ROS, which explains why no significant increase in antibacterial activity was observed compared to the control (CS) sample. On the other hand, curcumin provides antibacterial activity by settling on the cell membrane of bacteria and preventing the cell from dividing and multiplying [82], and as shown in Figure 4A, the curcumin-added suture showed 83.39% effectiveness in just 3 h and provided 98.87% bacterial inactivation at the end of 24 h.

The in vitro release profile of curcumin from CS/3T/C suture was analyzed at 37 °C in PBS (at pH 7.4) solution as a simulation of body fluid, and the cumulative release against incubation time is shown in Figure 4B. The curcumin-loaded suture showed a burst release of approximately 40% in the first two hours. Although different materials were used, a similar burst release phenomenon has been reported in the literature. For example, in the study by Rezaei and Nasirpour, curcumin-loaded almond gum/PVA nanofibers exhibited a release of approximately 27% within 2 h [83]. Rezaei and Nasirpour attributed this burst release to the weaker binding of curcumin on the fiber surface to the polymer and its higher diffusion tendency. Similarly, in this study, it is hypothesized that curcumin near the surface of the high surface area suture was released more rapidly into the simulated body fluid. Although the polymer matrices are different, the general mechanism of burst release by surface-bound curcumin may be comparable. The literature suggests that burst release assists wound healing by providing immediate relief followed by a sustained release phase [84,85]. The environment in which the bioactive substance is located, along with its interaction with the surroundings, plays a crucial role in its release performance. For instance, Bui et al. demonstrated that the addition of Tween 80 to PBS significantly affects the release of curcumin from Zein nanofibers. Their study showed that both the burst release and the total release amount increased as the Tween 80 concentration increased. Specifically, while the curcumin release was only 2.5% in PBS without Tween 80 after 72 h, it reached 11.5% in PBS containing 10% Tween 80 [86]. As depicted in Figure 4B, following the initial burst release within the first two hours, the sustained release of curcumin continued for up to 25 h, ultimately reaching a total release of approximately 77%. The curcumin release profile was consistent with the antibacterial test results. After three hours, curcumin release reached approximately 40%, corresponding to an 83% bacterial inactivation rate in the CS/3T/C suture. By the 12th hour, with a curcumin release of 62%, bacterial inactivation had reached nearly 99%. It is thought that burst release will have a positive effect on the healing of the injured area by cleaning the suture area from bacteria [87].

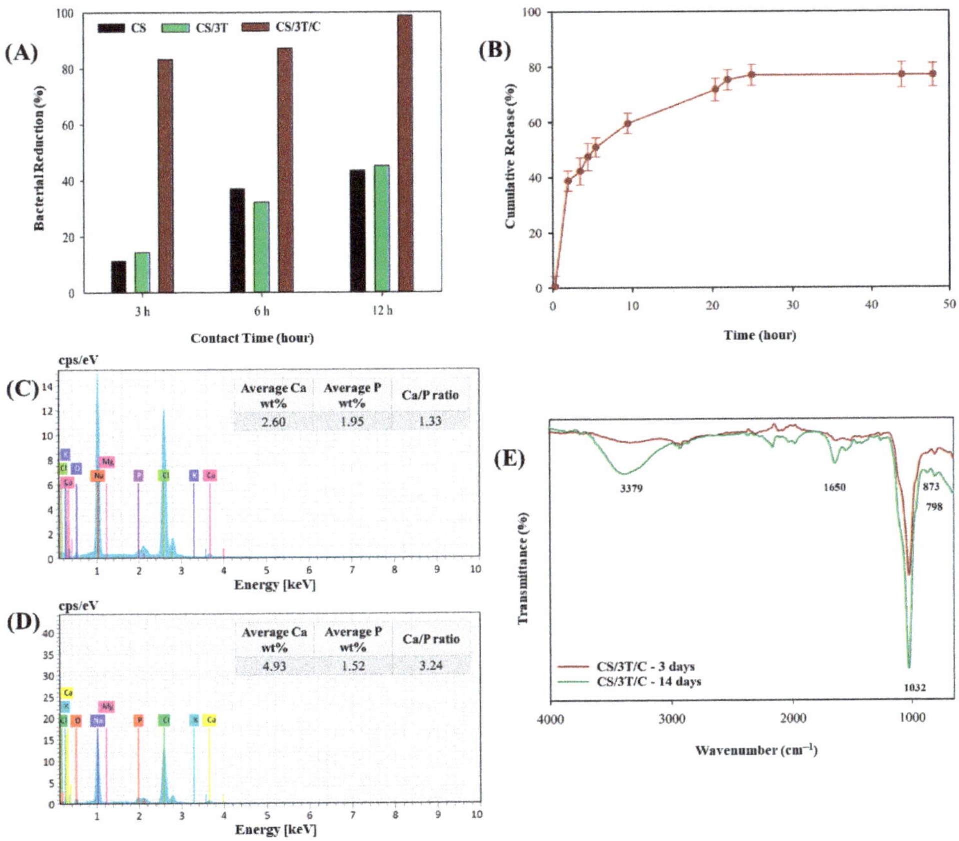

Figure 4. Antibacterial activity of CS, CS/3T, and CS/3T/C sutures against *S. aureus* at different time intervals (**A**), curcumin release profile of CS/3T/C suture sample (**B**), EDS spectra of the CS/3T/C suture after 3-day incubation in SBF (**C**), EDS spectra of the CS/3T/C suture after 14-day incubation in SBF (**D**), and FTIR spectra of the CS/3T/C suture after 3 and 14 days' incubation in SBF (**E**).

The hydroxyapatite layer formed on the surface of a material in contact with body fluids as a result of various biological reactions provides information about the bioactivity of the material [88,89]. In in vitro applications, 10 × SBF solution is used to accelerate the formation of this layer [90]. Figure 1 (After SBF) demonstrates bonelike crystalline hydroxyapatite layer formation on produced suture samples after 3 and 14 days of incubation. Granular core apatite formation in this layer showed a cauliflower-like morphology that is characteristic for bioactive surfaces, similar to our previous study [89]. The smooth surfaces of the sutures seen in Figure 1 (Before SBF) were covered with hydroxyapatite and completely change the morphology of the sutures. In addition, an increase in suture diameters was observed due to the accumulation of hydroxyapatite on their surfaces. Similar results were seen in the study by Boccaccini et al., who coated bioactive glass on polyglactin 910 sutures. It was observed that an approximately 15 µm HA layer was formed on the surface of the sutures after 21 days of incubation in SBF solution [91]. In addition to SEM analysis, EDS was used to determine the chemical composition of the cauliflower-like clusters on the CS/3T/C suture (Figure 4C,D). The Ca/P ratio on the surface of the CS/3T/C suture increased from 1.33 after 3 days of incubation to 3.24 after 14 days in SBF. The Ca and P elements, which were found to be concentrated in the nucleation regions, confirm that the formed layer on the scaffold consists of apatite.

Figure 4E showed the FTIR spectra of the chemical composition of the apatite deposit formed on the CS/3T/C suture after 3 and 14 days of incubation in SBF. The band at 798 cm^{-1} corresponded to the P-O bond stretching, indicating the presence of the crystalline phase of the apatite structure deposited on the surface of the CS/3T/C suture [92]. While the band at 1032 cm^{-1} was associated with the triply degenerated asymmetric stretching of PO_4^{3-}, the band around 873 cm^{-1} confirmed the presence of phosphate groups within the carbonated apatite structure [93,94]. On the other hand, the vibrations observed around 3379 cm^{-1} and 1650 cm^{-1}, which significantly increased from day 3 to day 14, corresponded to the O–H stretching of H_2O in the apatite layer [92–94].

Despite the reduced mechanical properties, the CS/3T/C suture demonstrated superior biological performance as seen in in vitro analysis. The incorporation of curcumin significantly improved antibacterial activity, bioactivity, and antioxidant performance, which are critical factors in reducing infection risks and accelerating wound healing. In surgical applications where infection control and bioactivity are prioritized over mechanical strength, such as in soft tissue repair or absorbable sutures, the benefits of curcumin-enhanced sutures may outweigh the mechanical limitations. Moreover, future studies could focus on optimizing the polymeric formulation or incorporating crosslinking strategies to enhance mechanical properties while preserving the biological advantages provided by curcumin.

4. Conclusions

In the present study, the dry jet–wet spinning method was successfully employed in developing smooth and homogeneous surface chitosan sutures with 7-0 suture number. The effects of the TiO_2 and curcumin content on the physicochemical, mechanical, morphological, antioxidant, and in vitro biocidal, drug release, and bioactivity properties of the sutures were systematically investigated. In the first part of the study, it was observed that the covalent bonds established between TiO_2 and chitosan provided an increase in the mechanical performance of chitosan sutures by up to 12.32% with 3% TiO_2 addition. While the addition of curcumin to increase the healing performance of the suture in the surgery area provided significant improvements in this sense, it caused a decrease in mechanical performance. In conclusion, curcumin-added chitosan sutures will contribute to the healing of the surgery area due to the improved antioxidant and biocidal activity and controlled release of the drug. To my knowledge, no previous studies have investigated curcumin-loaded TiO_2-reinforced chitosan surgical sutures. This pioneering work will pave the way for further research in this promising area and fills an important gap in the literature. By integrating innovative materials, this study not only expands the potential applications of advanced surgical sutures, but future studies can build on this foundation to improve their mechanical performance and explore broader biomedical applications.

Funding: This research received no external funding.

Institutional Review Board Statement: Not applicable.

Data Availability Statement: The original contributions presented in this study are included in the article. Further inquiries can be directed to the corresponding author.

Acknowledgments: I would like to thank Şeyma Duman and her team for generously providing the TiO_2 powder used in this study. I would also like to express my gratitude to Mehmet Orhan for his valuable support in conducting the bacterial tests.

Conflicts of Interest: The author declares no conflicts of interest.

References

1. Joseph, B.; James, J.; Kalarikkal, N.; Thomas, S. Advances in biopolymer based surgical sutures. In *Book Advanced Technologies and Polymer Materials for Surgical Sutures*, 1st ed.; Thomas, S., Coates, P., Whiteside, B., Joseph, B., Nair, K., Eds.; Woodhead Publishing: Cambridge, UK, 2023; pp. 1–10.
2. Öksüz, K.E.; Kurt, B.; İnan, Z.D.Ş.; Hepokur, C. Novel bioactive glass/graphene oxide-coated surgical sutures for soft tissue regeneration. *ACS Omega* **2023**, *8*, 21628–21641. [CrossRef]
3. Tan, Y.; Rajoka, M.S.R.; Ke, Z.; Mehwish, H.M.; Deng, W.; Li, J.; Qin, W.; Zhao, L.; Wu, Y. Effect of squid cartilage chitosan molecular structure on the properties of its monofilament as an absorbable surgical suture. *Polymers* **2022**, *14*, 1306. [CrossRef]
4. da Silva, M.C.; da Silva, H.N.; Alves Leal Cruz, R.D.C.; Sagoe Amoah, S.K.; de Lima Silva, S.M.; Lia Fook, M.V. N-acetyl-D-glucosamine-loaded chitosan filaments biodegradable and biocompatible for use as absorbable surgical suture materials. *Materials* **2019**, *12*, 1807. [CrossRef]
5. Li, H.; Cheng, F.; Chávez-Madero, C.; Choi, J.; Wei, X.; Yi, X.; Zheng, T.; He, J. Manufacturing and physical characterization of absorbable oxidized regenerated cellulose braided surgical sutures. *Int. J. Biol. Macromol.* **2019**, *134*, 56–62. [CrossRef] [PubMed]
6. Dennis, C.; Sethu, S.; Nayak, S.; Mohan, L.; Morsi, Y.; Manivasagam, G. Suture materials—Current and emerging trends. *J. Biomed. Mater. Res. Part A* **2016**, *104*, 1544–1599. [CrossRef] [PubMed]
7. Pillai, C.K.S.; Sharma, C.P. Absorbable polymeric surgical sutures: Chemistry, production, properties, biodegradability, and performance. *J. Biomater. Appl.* **2010**, *25*, 291–366. [CrossRef] [PubMed]
8. Barbosa, M.C.D.S.; Silva, H.N.D.; Lopes, D.D.S.; Wanderley, W.F.; Rosendo, R.A.; Penha, E.S.D.; de Medeiros, L.A.D.M.; Silva, S.M.d.L.; Fook, M.V.L. Biodegradable Chitosan Sutures Enhanced with N-Acetyl-D-Glucosamine: Comparative Study with Catgut Sutures. *Mater. Res.* **2024**, *27*, e20240278. [CrossRef]
9. Younesi, M.; Donmez, B.O.; Islam, A.; Akkus, O. Heparinized collagen sutures for sustained delivery of PDGF-BB: Delivery profile and effects on tendon-derived cells In-Vitro. *Acta Biomater.* **2016**, *41*, 100–109. [CrossRef]
10. Shao, K.; Han, B.; Gao, J.; Jiang, Z.; Liu, W.; Liu, W.; Liang, Y. Fabrication and feasibility study of an absorbable diacetyl chitin surgical suture for wound healing. *J. Biomed. Mater. Res. Part B Appl. Biomater.* **2016**, *104*, 116–125. [CrossRef] [PubMed]
11. Chu, C.C.; Williams, D.F. The effect of gamma irradiation on the enzymatic degradation of polyglycolic acid absorbable sutures. *J. Biomed. Mater. Res.* **1983**, *17*, 1029–1040. [CrossRef]
12. Im, J.N.; Kim, J.K.; Kim, H.K.; In, C.H.; Lee, K.Y.; Park, W.H. In vitro and in vivo degradation behaviors of synthetic absorbable bicomponent monofilament suture prepared with poly (p-dioxanone) and its copolymer. *Polym. Degrad. Stab.* **2007**, *92*, 667–674. [CrossRef]
13. Lim, T.Y.; Poh, C.K.; Wang, W. Poly (lactic-co-glycolic acid) as a controlled release delivery device. *J. Mater. Sci. Mater. Med.* **2009**, *20*, 1669–1675. [CrossRef] [PubMed]
14. Katz, A.R.; Mukherjee, D.P.; Kaganov, A.L.; Gordon, S. A new synthetic monofilament absorbable suture made from polytrimethylene carbonate. *Surg. Gynecol. Obstet.* **1985**, *161*, 213–222.
15. Hu, J.; Song, Y.; Zhang, C.; Huang, W.; Chen, A.; He, H.; Zhang, S.; Chen, Y.; Tu, C.; Liu, J.; et al. Highly aligned electrospun collagen/polycaprolactone surgical sutures with sustained release of growth factors for wound regeneration. *ACS Appl. Bio Mater.* **2020**, *3*, 965–976. [CrossRef]
16. Viju, S.; Thilagavathi, G. Effect of chitosan coating on the characteristics of silk-braided sutures. *J. Ind. Text.* **2013**, *42*, 256–268. [CrossRef]
17. Dash, M.; Chiellini, F.; Ottenbrite, R.M.; Chiellini, E. Chitosan—A versatile semi-synthetic polymer in biomedical applications. *Prog. Polym. Sci.* **2011**, *36*, 981–1014. [CrossRef]
18. Rodrigues, S.; Dionísio, M.; Remunan Lopez, C.; Grenha, A. Biocompatibility of chitosan carriers with application in drug delivery. *J. Funct. Biomater.* **2012**, *3*, 615–641. [CrossRef]
19. Marțău, G.A.; Mihai, M.; Vodnar, D.C. The use of chitosan, alginate, and pectin in the biomedical and food sector—Biocompatibility, bioadhesiveness, and biodegradability. *Polymers* **2019**, *11*, 1837. [CrossRef] [PubMed]
20. Rizeq, B.R.; Younes, N.N.; Rasool, K.; Nasrallah, G.K. Synthesis, bioapplications, and toxicity evaluation of chitosan-based nanoparticles. *Int. J. Mol. Sci.* **2019**, *20*, 5776. [CrossRef] [PubMed]
21. Tan, H.; Ma, R.; Lin, C.; Liu, Z.; Tang, T. Quaternized chitosan as an antimicrobial agent: Antimicrobial activity, mechanism of action and biomedical applications in orthopedics. *Int. J. Mol. Sci.* **2013**, *14*, 1854–1869. [CrossRef]
22. Panda, P.K.; Sadeghi, K.; Park, K.; Seo, J. Regeneration approach to enhance the antimicrobial and antioxidant activities of chitosan for biomedical applications. *Polymers* **2022**, *15*, 132. [CrossRef] [PubMed]
23. Khalil, K.D.; Bashal, A.H.; Habeeb, T.; Abu-Dief, A.M. Synergistic antibacterial and anticancer activity in gadolinium–chitosan nanocomposite films: A novel approach for biomedical applications. *Appl. Organomet. Chem.* **2024**, *38*, e7531. [CrossRef]
24. Matica, M.A.; Aachmann, F.L.; Tøndervik, A.; Sletta, H.; Ostafe, V. Chitosan as a wound dressing starting material: Antimicrobial properties and mode of action. *Int. J. Mol. Sci.* **2019**, *20*, 5889. [CrossRef]

25. Montenegro, R.; Godeiro, J.R.G. Chitosan based suture–focusing on the real advantages of an outstanding biomaterial. *Adv. Chitin Sci.* **2014**, *14*, 211–216.
26. Perrin, N.; Mohammadkhani, G.; Moghadam, F.H.; Delattre, C.; Zamani, A. Biocompatible fibers from fungal and shrimp chitosans for suture application. *Curr. Res. Biotechnol.* **2022**, *4*, 530–536. [CrossRef]
27. Farrokhi, H.; Koosha, M.; Nasirizadeh, N.; Salari, M.; Abdouss, M.; Li, T.; Gong, Y. The Effect of Nanoclay Type on the Mechanical Properties and Antibacterial Activity of Chitosan/PVA Nanocomposite Films. *J. Compos. Sci.* **2024**, *8*, 255. [CrossRef]
28. Aryaei, A.; Jayatissa, A.H.; Jayasuriya, A.C. Mechanical and biological properties of chitosan/carbon nanotube nanocomposite films. *J. Biomed. Mater. Res. Part A* **2014**, *102*, 2704–2712. [CrossRef] [PubMed]
29. Marroquin, J.B.; Rhee, K.Y.; Park, S.J. Chitosan nanocomposite films: Enhanced electrical conductivity, thermal stability, and mechanical properties. *Carbohydr. Polym.* **2013**, *92*, 1783–1791. [CrossRef] [PubMed]
30. Mahalakshmi, S.; Vijaya, P. Evaluation of in-vitro biocompatibility and antimicrobial activities of titanium dioxide (TiO_2) nanoparticles by hydrothermal method. *Nano Biomed. Eng.* **2021**, *13*, 36–43. [CrossRef]
31. Kiran, A.S.K.; Kumar, T.S.; Sanghavi, R.; Doble, M.; Ramakrishna, S. Antibacterial and bioactive surface modifications of titanium implants by PCL/TiO_2 nanocomposite coatings. *Nanomaterials* **2018**, *8*, 860. [CrossRef]
32. Khan, S.; Garg, M.; Chockalingam, S.; Gopinath, P.; Kundu, P.P. TiO_2 doped chitosan/poly (vinyl alcohol) nanocomposite film with enhanced mechanical properties for application in bone tissue regeneration. *Int. J. Biol. Macromol.* **2020**, *143*, 285–296. [CrossRef] [PubMed]
33. Cano, L.; Pollet, E.; Avérous, L.; Tercjak, A. Effect of TiO_2 nanoparticles on the properties of thermoplastic chitosan-based nano-biocomposites obtained by mechanical kneading. *Compos. Part A Appl. Sci. Manuf.* **2017**, *93*, 33–40. [CrossRef]
34. Ravandi, R.; Heris, S.Z.; Hemmati, S.; Aghazadeh, M.; Davaran, S.; Abdyazdani, N. Effects of chitosan and TiO_2 nanoparticles on the antibacterial property and ability to self-healing of cracks and retrieve mechanical characteristics of dental composites. *Heliyon* **2024**, *10*, e27734. [CrossRef] [PubMed]
35. Prajapati, S.K.; Malaiya, A.; Kesharwani, P.; Soni, D.; Jain, A. Biomedical applications and toxicities of carbon nanotubes. *Drug Chem. Toxicol.* **2022**, *45*, 435–450. [CrossRef] [PubMed]
36. Merino, D.; Tomadoni, B.; Salcedo, M.F.; Mansilla, A.Y.; Casalongué, C.A.; Alvarez, V.A. Nanoclay as carriers of bioactive molecules applied to agriculture. In *Handbook of Nanomaterials and Nanocomposites for Energy and Environmental Applications*, 1st ed.; Kharissova, O.V., Torres-Martínez, L.M., Kharisov, B.I., Eds.; Springer: Cham, Switzerland, 2022; pp. 1–22.
37. Jeon, J.D.; Kim, M.J.; Kwak, S.Y. Effects of addition of TiO_2 nanoparticles on mechanical properties and ionic conductivity of solvent-free polymer electrolytes based on porous P (VdF-HFP)/P (EO-EC) membranes. *J. Power Sources* **2006**, *162*, 1304–1311. [CrossRef]
38. Qu, L.; Chen, G.; Dong, S.; Huo, Y.; Yin, Z.; Li, S.; Chen, Y. Improved mechanical and antimicrobial properties of zein/chitosan films by adding highly dispersed nano-TiO_2. *Ind. Crops Prod.* **2019**, *130*, 450–458. [CrossRef]
39. El-Wakil, N.A.; Hassan, E.A.; Abou-Zeid, R.E.; Dufresne, A. Development of wheat gluten/nanocellulose/titanium dioxide nanocomposites for active food packaging. *Carbohydr. Polym.* **2015**, *124*, 337–346. [CrossRef] [PubMed]
40. Nocito, M.C.; De Luca, A.; Prestia, F.; Avena, P.; La Padula, D.; Zavaglia, L.; Sirianni, R.; Casaburi, I.; Puoci, F.; Chimento, A.; et al. Antitumoral activities of curcumin and recent advances to improve its oral bioavailability. *Biomedicines* **2021**, *9*, 1476. [CrossRef] [PubMed]
41. Zheng, D.; Huang, C.; Huang, H.; Zhao, Y.; Khan, M.R.U.; Zhao, H.; Huang, L. Antibacterial mechanism of curcumin: A review. *Chem. Biodivers.* **2020**, *17*, e2000171. [CrossRef]
42. Chen, J.; He, Z.M.; Wang, F.L.; Zhang, Z.S.; Liu, X.Z.; Zhai, D.D.; Chen, W.D. Curcumin and its promise as an anticancer drug: An analysis of its anticancer and antifungal effects in cancer and associated complications from invasive fungal infections. *Eur. J. Pharmacol.* **2016**, *772*, 33–42. [CrossRef]
43. Jennings, M.R.; Parks, R.J. Curcumin as an antiviral agent. *Viruses* **2020**, *12*, 1242. [CrossRef] [PubMed]
44. Menon, V.P.; Sudheer, A.R. Antioxidant and anti-inflammatory properties of curcumin. In *The Molecular Targets and Therapeutic Uses of Curcumin in Health and Disease*, 1st ed; Aggarwal, B.B., Surh, Y., Shishodia, S., Eds.; Springer Science & Business Media: New York, NY, USA, 2007; pp. 105–125.
45. Kim, D.C.; Ku, S.K.; Bae, J.S. Anticoagulant activities of curcumin and its derivative. *BMB Rep.* **2012**, *45*, 221–226. [CrossRef]
46. Riyad, P.; Purohit, A.; Karishma, S.; Ram, H. Atherosclerotic plaque regression and HMG-CoA reductase inhibition potential of curcumin: An integrative omics and in-vivo study. *J. Appl. Biol. Biotechnol.* **2022**, *10*, 129–135.
47. Park, J.; Conteas, C.N. Anti-carcinogenic properties of curcumin on colorectal cancer. *World J. Gastrointest. Oncol.* **2010**, *2*, 169. [CrossRef]
48. Puneeth, H.R.; Sharada, A.C. Antioxidant and hypoglycemic effects of curcumin pyrazole derivatives. *Int. J. Pharm. Pharm. Sci.* **2015**, *7*, 244–249.
49. Bulut, B.; Duman, Ş. Effects of calcination temperature on hydrothermally synthesized titanium dioxide submicron powders. *Konya J. Eng. Sci.* **2021**, *9*, 676–685. [CrossRef]

50. Yudin, V.E.; Dobrovolskaya, I.P.; Neelov, I.M.; Dresvyanina, E.N.; Popryadukhin, P.V.; Ivan'kova, E.M.; Elokhovskii, V.Y.; Kasatkin, I.A.; Okrugin, B.M.; Morganti, P. Wet spinning of fibers made of chitosan and chitin nanofibrils. *Carbohydr. Polym.* **2014**, *108*, 176–182. [CrossRef] [PubMed]
51. Mohammadkhani, G.; Kumar Ramamoorthy, S.; Adolfsson, K.H.; Mahboubi, A.; Hakkarainen, M.; Zamani, A. New solvent and coagulating agent for development of chitosan fibers by wet spinning. *Polymers* **2021**, *13*, 2121. [CrossRef] [PubMed]
52. Dresvyanina, E.N.; Dobrovol'Skaya, I.P.; Popryadukhin, P.V.; Yudin, V.E.; Ivan'Kova, E.M.; Elokhovskii, V.Y.; Khomenko, A.Y. Influence of spinning conditions on properties of chitosan fibers. *Fibre Chem.* **2013**, *44*, 280–283. [CrossRef]
53. *ASTM D2256/D2256M-21*; Standard Test Method for Tensile Properties of Yarns by the Single-Strand Method. ASTM: West Conshohocken, PA, USA, 2022.
54. Mavis, B.; Demirtaş, T.T.; Gümüşderelioğlu, M.; Gündüz, G.; Çolak, Ü. Synthesis, characterization and osteoblastic activity of polycaprolactone nanofibers coated with biomimetic calcium phosphate. *Acta Biomater.* **2009**, *5*, 3098–3111. [CrossRef]
55. Hussein, E.M.; Desoky, W.M.; Hanafy, M.F.; Guirguis, O.W. Effect of TiO_2 nanoparticles on the structural configurations and thermal, mechanical, and optical properties of chitosan/TiO_2 nanoparticle composites. *J. Phys. Chem. Solids* **2021**, *152*, 109983. [CrossRef]
56. Knaul, J.; Hooper, M.; Chanyi, C.; Creber, K.A. Improvements in the drying process for wet-spun chitosan fibers. *J. Appl. Polym. Sci.* **1998**, *69*, 1435–1444. [CrossRef]
57. Nechyporchuk, O.; Yang Nilsson, T.; Ulmefors, H.; Köhnke, T. Wet spinning of chitosan fibers: Effect of sodium dodecyl sulfate adsorption and enhanced dope temperature. *ACS Appl. Polym. Mater.* **2020**, *2*, 3867–3875. [CrossRef]
58. Pantić, M.; Maver, U.; Rožanc, J.; Vihar, B.; Andrejč, D.C.; Knez, Ž.; Novak, Z. Evaluation of ethanol-induced chitosan aerogels with human osteoblast cells. *Int. J. Biol. Macromol.* **2023**, *253*, 126694. [CrossRef] [PubMed]
59. Queiroz, M.F.; Teodosio Melo, K.R.; Sabry, D.A.; Sassaki, G.L.; Rocha, H.A.O. Does the use of chitosan contribute to oxalate kidney stone formation? *Mar. Drugs* **2014**, *13*, 141–158. [CrossRef] [PubMed]
60. Ibekwe, C.A.; Oyatogun, G.M.; Esan, T.A.; Oluwasegun, K.M. Synthesis and characterization of chitosan/gum arabic nanoparticles for bone regeneration. *Am. J. Mater. Sci. Eng.* **2017**, *5*, 28–36.
61. Rachtanapun, P.; Klunklin, W.; Jantrawut, P.; Jantanasakulwong, K.; Phimolsiripol, Y.; Seesuriyachan, P.; Leksawasdi, N.; Chaiyaso, T.; Ruksiriwanich, W.; Phongthai, S.; et al. Characterization of chitosan film incorporated with curcumin extract. *Polymers* **2021**, *13*, 963. [CrossRef] [PubMed]
62. Karthikeyan, K.T.; Nithya, A.; Jothivenkatachalam, K. Photocatalytic and antimicrobial activities of chitosan-TiO_2 nanocomposite. *Int. J. Biol. Macromol.* **2017**, *104*, 1762–1773. [CrossRef] [PubMed]
63. Musielak, E.; Feliczak-Guzik, A.; Jaroniec, M.; Nowak, I. Photodynamic light-triggered release of curcumin from hierarchical FAU zeolite. *Catalysts* **2023**, *13*, 394. [CrossRef]
64. Notin, L.; Viton, C.; David, L.; Alcouffe, P.; Rochas, C.; Domard, A. Morphology and mechanical properties of chitosan fibers obtained by gel-spinning: Influence of the dry-jet-stretching step and ageing. *Acta Biomater.* **2006**, *2*, 387–402. [CrossRef]
65. Albanna, M.Z.; Bou-Akl, T.H.; Walters, H.L., III; Matthew, H.W. Improving the mechanical properties of chitosan-based heart valve scaffolds using chitosan fibers. *J. Mech. Behav. Biomed. Mater.* **2012**, *5*, 171–180. [CrossRef] [PubMed]
66. Ciftci, F.; Özarslan, A.C.; Evcimen Duygulu, N. Production and comprehensive characterization of PVA/chitosan transdermal composite mats loaded with bioactive curcumin; evaluation of its release kinetics, antioxidant, antimicrobial, and biocompatibility features. *J. Appl. Polym. Sci.* **2024**, *141*, e55874. [CrossRef]
67. Abd El-Hady, M.M.; Saeed, S.E.S. Antibacterial properties and pH sensitive swelling of insitu formed silver-curcumin nanocomposite based chitosan hydrogel. *Polymers* **2020**, *12*, 2451. [CrossRef]
68. Pham-Huy, L.A.; He, H.; Pham-Huy, C. Free radicals, antioxidants in disease and health. *Int. J. Biomed. Sci. IJBS* **2008**, *4*, 89. [CrossRef]
69. Yuan, Y.; Tan, W.; Zhang, J.; Li, Q.; Guo, Z. Water-soluble amino functionalized chitosan: Preparation, characterization, antioxidant and antibacterial activities. *Int. J. Biol. Macromol.* **2022**, *217*, 969–978. [CrossRef]
70. Rasheed, U.; Kiani, M.N.; Butt, M.S.; Saeed, H.; Hanif, R.; Anwar, S. Fabrication and biocompatibility of neem/chitosan coated silk sutures for infection control and wound healing. *J. King Saud Univ. Sci.* **2024**, *36*, 103435. [CrossRef]
71. Shanmugam, A.; Subhapradha, N.; Suman, S.; Ramasamy, P.; Saravanan, R.; Shanmugam, V.; Srinivasan, A. Characterization of biopolymer "chitosan" from the shell of donacid clam *Donax scortum* (Linnaeus, 1758) and its antioxidant activity. *Int. J. Pharm. Pharm. Sci.* **2012**, *4*, 460–465.
72. Si Trung, T.; Bao, H.N.D. Physicochemical properties and antioxidant activity of chitin and chitosan prepared from pacific white shrimp waste. *Int. J. Carbohydr. Chem.* **2015**, *2015*, 706259. [CrossRef]
73. Xia, W.; Liu, P.; Zhang, J.; Chen, J. Biological activities of chitosan and chitooligosaccharides. *Food Hydrocoll.* **2011**, *25*, 170–179. [CrossRef]
74. Zhang, X.; Liu, Y.; Yong, H.; Qin, Y.; Liu, J.; Liu, J. Development of multifunctional food packaging films based on chitosan, TiO_2 nanoparticles and anthocyanin-rich black plum peel extract. *Food Hydrocoll.* **2019**, *94*, 80–92. [CrossRef]

75. Ullattil, S.G.; Narendranath, S.B.; Pillai, S.C.; Periyat, P. Black TiO_2 nanomaterials: A review of recent advances. *Chem. Eng. J.* **2018**, *343*, 708–736. [CrossRef]
76. Kazan, A.; Demirci, F. Olive leaf extract incorporated chitosan films for active food packaging. *Konya J. Eng. Sci.* **2023**, *11*, 1061–1072. [CrossRef]
77. Bi, F.; Qin, Y.; Chen, D.; Kan, J.; Liu, J. Development of active packaging films based on chitosan and nano-encapsulated luteolin. *Int. J. Biol. Macromol.* **2021**, *182*, 545–553. [CrossRef] [PubMed]
78. Zhang, Q.; Qiao, Y.; Li, C.; Lin, J.; Han, H.; Li, X.; Mao, J.; Wang, F.; Wang, L. Chitosan/gelatin-tannic acid decorated porous tape suture with multifunctionality for tendon healing. *Carbohydr. Polym.* **2021**, *268*, 118246. [CrossRef] [PubMed]
79. Zhang, Q.; Qiao, Y.; Zhu, J.; Li, Y.; Li, C.; Lin, J.; Li, X.; Han, H.; Mao, J.; Wang, F.; et al. Electroactive and antibacterial surgical sutures based on chitosan-gelatin/tannic acid/polypyrrole composite coating. *Compos. Part B Eng.* **2021**, *223*, 109140. [CrossRef]
80. Zhang, X.; Xiao, G.; Wang, Y.; Zhao, Y.; Su, H.; Tan, T. Preparation of chitosan-TiO_2 composite film with efficient antimicrobial activities under visible light for food packaging applications. *Carbohydr. Polym.* **2017**, *169*, 101–107. [CrossRef] [PubMed]
81. Sagadevan, S.; Lett, J.A.; Vennila, S.; Prasath, P.V.; Kaliaraj, G.S.; Fatimah, I.; Léonard, E.; Mohammad, F.; Al-Lohedan, H.A.; Alshahateet, S.F.; et al. Photocatalytic activity and antibacterial efficacy of titanium dioxide nanoparticles mediated by Myristica fragrans seed extract. *Chem. Phys. Lett.* **2021**, *771*, 138527. [CrossRef]
82. Li, H.; Jiang, Y.; Yang, J.; Pang, R.; Chen, Y.; Mo, L.; Jiang, Q.; Qin, Z. Preparation of curcumin-chitosan composite film with high antioxidant and antibacterial capacity: Improving the solubility of curcumin by encapsulation of biopolymers. *Food Hydrocoll.* **2023**, *145*, 109150. [CrossRef]
83. Rezaei, A.; Nasirpour, A. Evaluation of release kinetics and mechanisms of curcumin and curcumin-β-cyclodextrin inclusion complex incorporated in electrospun almond gum/PVA nanofibers in simulated saliva and simulated gastrointestinal conditions. *BioNanoScience* **2019**, *9*, 438–445. [CrossRef]
84. Cam, M.E.; Yildiz, S.; Alenezi, H.; Cesur, S.; Ozcan, G.S.; Erdemir, G.; Edirisinghe, U.; Akakin, D.; Kuruca, D.S.; Kabasakal, L.; et al. Evaluation of burst release and sustained release of pioglitazone-loaded fibrous mats on diabetic wound healing: An in vitro and in vivo comparison study. *J. R. Soc. Interface* **2020**, *17*, 20190712. [CrossRef]
85. Huang, X.; Brazel, C.S. On the importance and mechanisms of burst release in matrix-controlled drug delivery systems. *J. Control. Release* **2001**, *73*, 121–136. [CrossRef]
86. Bui, H.T.; Chung, O.H.; Park, J.S. Fabrication of electrospun antibacterial curcumin-loaded zein nanofibers. *Polymer* **2014**, *38*, 744–751. [CrossRef]
87. Houshyar, S.; Yin, H.; Pope, L.; Zizhou, R.; Dekiwadia, C.; Hill-Yardin, E.L.; MC Yeung, J.; John, S.; Fox, K.; Tran, N.; et al. Smart suture with iodine contrasting nanoparticles for computed tomography. *OpenNano* **2023**, *9*, 100120. [CrossRef]
88. Blaker, J.J.; Nazhat, S.N.; Boccaccini, A.R. Development and characterisation of silver-doped bioactive glass-coated sutures for tissue engineering and wound healing applications. *Biomaterials* **2004**, *25*, 1319–1329. [CrossRef] [PubMed]
89. Mutlu, B.; Demirci, F.; Duman, Ş. Investigating the impact of coagulation bath temperature on the properties of biphasic calcium phosphate/poly (vinylidene fluoride)-based membrane scaffold via immersion precipitation. *J. Appl. Polym. Sci.* **2024**, *142*, e56375. [CrossRef]
90. Duman, Ş.; Bulut, B. Effect of akermanite powders on mechanical properties and bioactivity of chitosan-based scaffolds produced by 3D-bioprinting. *Ceram. Int.* **2021**, *47*, 13912–13921. [CrossRef]
91. Boccaccini, A.R.; Stamboulis, A.G.; Rashid, A.; Roether, J.A. Composite surgical sutures with bioactive glass coating. *J. Biomed. Mater. Res. Part B Appl. Biomater.* **2003**, *67*, 618–626. [CrossRef]
92. Sierra, L.A.Q.; Escobar, D.M. Characterization and bioactivity behavior of sol–gel derived bioactive vitroceramic from non-conventional precursors. *Boletín De La Soc. Española De Cerámica Y Vidr.* **2019**, *58*, 85–92. [CrossRef]
93. Wei, D.; Zhou, Y.; Jia, D.; Wang, Y. Biomimetic apatite deposited on microarc oxidized anatase-based ceramic coating. *Ceram. Int.* **2008**, *34*, 1139–1144. [CrossRef]
94. Mutlu, B.; Demirci, F.; Duman, Ş. Influence of boron incorporated biphasic calcium phosphate on mechanical, thermal, and biological properties of poly (vinylidene fluoride) membrane scaffold. *J. Am. Ceram. Soc.* **2024**, *107*, 7274–7288. [CrossRef]

polymers

MDPI

Article

Evaluation of Mechanical Properties of Composite Material with a Thermoplastic Matrix Reinforced with Cellulose Acetate Microfibers

Pedro Rodríguez Sandoval [1], Andres Felipe Rubiano-Navarrete [2,*], Edwin Yesid Gómez-Pachón [3] and Ricardo Vera-Graziano [4]

[1] Grupo de Investigación de Materiales y Ensayos—GIMES, SENA—Centro de Materiales y Ensayos, Escuela de Posgrado en Ingeniería, Universidad Pedagógica y Tecnológica de Colombia, UPTC, Tunja 150003, Colombia; pedro.rodriguez01@uptc.edu.co

[2] Grupo de Investigación en Diseño, Innovación y Asistencia Técnica para Materiales Avanzados-DITMAV, Maestría en Metalurgia y Ciencia de los Materiales, Universidad Pedagógica y Tecnológica de Colombia-UPTC, Tunja 150003, Colombia

[3] Grupo de Investigación en Diseño, Innovación y Asistencia Técnica para Materiales Avanzados-DITMAV, Escuela de Diseño Industrial, Universidad Pedagógica y Tecnológica de Colombia-UPTC, Duitama 150461, Colombia; edwin.gomez02@uptc.edu.co

[4] Instituto de Investigaciones en Materiales, Universidad Nacional Autónoma de México, Mexico City 04510, Mexico

* Correspondence: andres.rubiano01@uptc.edu.co; Tel.: +57-314-27-99-039

Abstract: Low-density polyethylene (LDPE) has been widely used in various applications due to its flexibility, lightness, and low production cost. However, its massive use in disposable products has raised environmental concerns, prompting the search for more sustainable alternatives. This study aims to investigate the mechanical properties achievable in a composite material utilizing low-density polyethylene (LDPE), potato starch (PS), and cellulose microfibrils (MFCA) at loadings of 0.05%, 0.15%, and 0.30%. Initially, the cellulose acetate microfibrils (MFCA) were produced via an electrospinning process. Subsequently, a dispersive mixture of the aforementioned materials was created through the extrusion and pelletizing process to form pellets. These pellets were then molded by injection molding to produce test specimens in accordance with ASTM D 638, the standard for tensile strength testing. The evaluation of the properties was conducted through mechanical tensile tests (ASTM D638), hardness tests (ASTM D 2240), melt flow index (ASTM D1238), and scanning electron microscopy (SEM). This study determined the influence of cellulose acetate microfibril loadings below 0.3% as reinforcement within a thermoplastic LDPE matrix. It was demonstrated that these microfibrils, due to their length-to-diameter ratio, contribute to an enhancement in the mechanical properties.

Keywords: extrusion; low-density polyethylene; microfibers; polymer; reinforcement

1. Introduction

Composite materials have become a formidable alternative across various industrial sectors, which is attributable to their high mechanical strength, stiffness, thermal resistance, and hardness, among other properties. These composites are characterized by a blend of materials that achieve emergent properties necessary for specific applications [1,2]. Recently, research has concentrated on developing composite materials endowed with the ability to degrade over time, as a result of interaction with organic matter and microbial agents. The majority of these materials are formulated using a synthetic polymer matrix reinforced with natural fibers or particles [3,4].

The pursuit of the ideal material has led to experiments with reinforcements or fillers that enhance physical properties without compromising their degradability. Consequently, there has been a pivot towards using natural fibers due to their low cost [5–8], ready availability, and the benefit of utilizing natural waste. Some commonly used fibers for fillers or reinforcements are those derived from coconut, fique, hemp, cocoa, among others. This is a result of the amalgamation of mechanical properties, stiffness, low thermal expansion coefficient, and chemical inertia [9].

The incorporation of microfibrils as reinforcements in materials has significantly contributed to advancements in the creation of materials, as research has focused on micrometric scale reinforcements to ensure better compatibility with synthetic matrices and natural polymers. Various types of micrometric reinforcements exist; however, those of a cellulosic nature have been the most extensively studied and utilized due to cellulose being one of the most abundant structural polymers on earth. Additionally, they exhibit high tensile strength, stiffness, flexibility, and superior mechanical, electrical, thermal, and dynamic properties compared to other commercial fibers [10,11], which render them an ecologically sound reinforcement alternative.

Microfibril acquisition is accomplished through various methods, one of which is electrospinning. This technique involves connecting a polymer solution, contained within a syringe with a metallic needle, to a high-voltage power source where an electrical charge is applied (positive pole), with a metallic collector (negative pole) situated at the other end, typically made of aluminum or copper. This is where the microfibrils are deposited [6].

The electrospinning process begins when voltage is applied at the tip of the needle, creating a droplet in the shape of a cone, which is a consequence of its electrostatic polarization [12]. Once the electric field's force exceeds the surface tension of the polymer solution, it is propelled toward the collector in the form of a filament. En route to the collector, the solvent evaporates, leading to the formation of a microfibril that is deposited on the collector, subsequently forming a nonwoven membrane [13].

To integrate microcompounds into matrices, a transformation process is necessary to facilitate their incorporation. This has been achieved using various polymer transformation processes such as injection molding, compression, thermoforming, and extrusion, with the latter being the most widely utilized [14–16]. Extrusion is a high-temperature, short-time process involving thermal, pressure, and mechanical shearing [17]. This continuous process allows for product quality control by adjusting the temperature and the screw speed [18], making it a versatile transformation process applicable across different industrial sectors.

In this study, an examination of the mechanical properties of a composite material composed of PE/starch reinforced with cellulose microfibers obtained via electrospinning was conducted. The composite material was produced by the extrusion process using a twin-screw extruder for this purpose.

2. Materials and Methods

2.1. Materials

For the preparation of the composite material, low-density polyethylene (LDPE) from ECOPETROL, Bogotá, Colombia (this low-density polymer is a synthetic polymer derived from petroleum), reference POLIFÉN 641; modified potato starch (PS) from EMSLAND GROUP, potato starch reference; Polyethylene glycol—PG, from CHEMICOL CH S.A.S, reference 6000—batch 200115W001203; and distilled water from CEPROSA, reference distilled, were employed.

The cellulose acetate microfibers (MFCA) were manufactured using the following materials: Cellulose acetate from SIGMA ALDRICH, reference 419028; Acetone from SIGMA ALDRICH, reference 179973-4L; Ethanol from EMSURE, reference 100983-100; and Chloroform from J.T. Baker, reference 918003-4L.

2.2. Preparation of Raw Materials

Electrospinning Process

For the preparation of the electrospinning material solution, an exploratory experimental design was conducted that included variations in polymer concentrations and the solvent mixture ratio. As a result of various exploratory tests, a standard concentration of 6.5% cellulose acetate polymer and a solvent ratio of 2:1 Acetone/Chloroform were established.

The laboratory equipment used included digital scales (brands: RADWAG, reference: ac 220/C/2; AVANPRO, reference: ABS), magnetic stirrers (brands: STABLETEMP, reference: COLE-PAEMER; TENAL, reference: TE-0854), 50 mL beakers, graduated cylinders, universal stands, watch glasses, glass jars with plastic lids, and magnets. The cellulose acetate microfibers (MFCA) were obtained by the electrospinning process using a high-voltage power supply (HV350R), injection pump systems Inc NE-1000 (Farmingdale, NY, USA), a copper collector covered with aluminum, a 10 mL Plasticak brand plastic syringe, and a steel needle with a capillary diameter of 0.8 mm developed and constructed by the DITMAV research group of the Universidad Pedagógica y Tecnológica de Colombia UPTC.

For this process, the variables for obtaining the microfibers were established, taking into account previous research, and the process was verified for the homogeneous membrane formation by setting the parameters to a flow rate of 30 mL/h, a distance of 30 cm, and a voltage of 17 KV, as shown in Table 1 [14]. Once the membrane was obtained, a micrograph was taken and the fiber diameter was measured using ImageJ (version 64-bit Java 8) software.

Table 1. Standardization of electrospinning process parameters.

Parameters	Values
Solution flow rate	30 mL/hra
Voltage	17 kV
Distance between needle and collector	30 cm
Volume	5 mL
Polymer concentration	6%

Extrusion/Pelletizing Process

The production of the composite material, composed of a thermoplastic matrix of LDPE/PS reinforced with cellulose acetate microfibers, was conducted at the SENA—Materials and Testing Laboratory.

Subsequently, the measurement of the mass of LDPE, PS, and MFCA commenced, bearing in mind that PS should constitute 15% by weight of the mixture and the proportion of LDPE is directly dependent on the MFCA percentages of 0.05%, 0.15%, and 0.30% of the total mixture weight, resulting in the treatments outlined in Table 2.

Table 2. Experimental Design Treatments.

Treatments			
T	Mixtures (%)		
	LDPE	PS	MFCA
T_1	84.95	15	0.05
T_2	84.85	15	0.15
T_3	84.7	15	0.30
T_4	85	15	0.00

The materials were integrated using a SHINI brand vertical-type industrial mixing hopper at a speed of 120 rpm for 10 min, after which they were packaged in bags labeled according to the treatments of the experimental design.

This process began with the setup and programming of the extrusion/pelletizing line in accordance with the treatments listed in Table 3, which details process variables such as temperatures, pressures, speeds, screw rpm, and cycle times.

Table 3. Extrusion/pelletizing parameter sheet.

Process Parameters: Extrusion/Pelletizing	
Temperatures	Zone 1: 50 °C
	Zone 2: 80 °C
	Zone 3: 100 °C
	Zone 4: 140 °C
	Zone 5: 165 °C
Cooling Tank Temperature	10 °C
Extruder Screw Speed	30 RPM
Pelletizer Speed	40 RPM
Extruder Motor Frequency	50 Hz
Pelletizer Motor Frequency	60 Hz

A double-screw extruder, series 14198 reference PTL-30, was utilized, which was programmed with a heating curve ranging from 50 to 165 °C across 5 zones and a screw rotation speed of 40 rpm. The pre-heated mixture at 50 °C was added to the machine's feed hopper and passed through the cylinder, with the screws conveying it to the head where it exits through four dies forming plastic filaments. These filaments then pass through a cooling tank where water under pressure circulates in a closed loop to solidify the material, which is subsequently transported to the pelletizing equipment to form pellets of 4 mm in diameter and 4 mm in length, as depicted in Figure 1. From each treatment outlined in Table 3, 500 g were obtained to prepare the samples and perform physical property characterization tests.

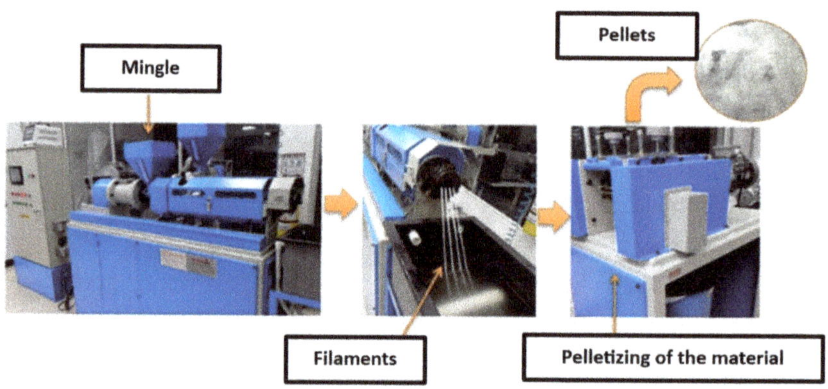

Figure 1. Extrusion/pelletizing process for reinforced material.

This extrusion/pelletizing process was parameterized according to works on the extrusion of plastics by Rodriguez, P., Prieto, E., Pachón, Palange, C., Marcus, A., Scurr, D., Stephen, P., Eichhorn, J. [16,19].

Injection Molding of Test Specimens.

For the production of test specimens in accordance with ASTM D-638 [20], used for tensile mechanical tests, the injection molding process was utilized with a two-cavity cold runner and automatic ejection metal mold. The materials developed with each of the established treatments were preconditioned in a SHINI brand pre-dryer set to a temperature of 80 °C for 15 min to remove moisture. They were then loaded into the hopper of the injection machine, where they gravitated into the screw chamber and were homogenized and plasticized before being injected into the mold, as shown in Figure 2.

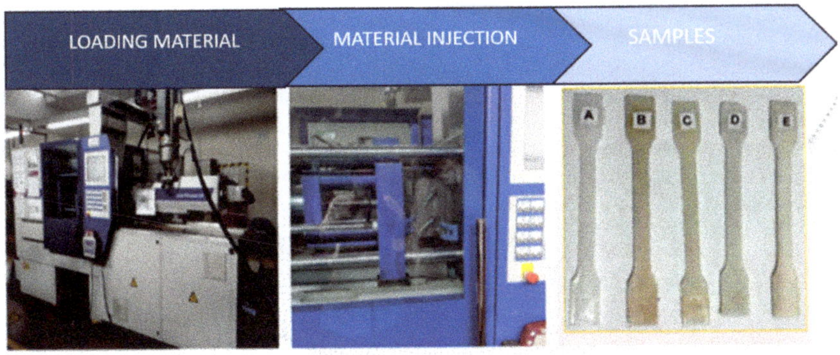

Figure 2. Injection molding process of test specimens. (A) LDPE 100%, PS 0.0%, MFCA 0. 0%; (B) LDPE 84.95%, PS 15%, MFCA 0. 05%; (C) LDPE 84.85%, PS 15%, MFCA 0.15%; (D) LDPE 84.7%, PS 15%, MFCA 0.30%; (E) LDPE 85%, PS 15%, MFCA 0.00%.

When the material is injected into the mold, it is cooled by water circulation within the mold tailored to the geometry of the part to ensure rapid solidification. The mold then opens, and the part is ejected by an ejection system and retrieved by a robotic arm that places it on a conveyor belt for final collection.

The injection molding process for the potato starch with low-density polyethylene was programmed according to the parameters detailed in the research work of Rodríguez, P., Prieto, E., Pachón, Y. [16].

2.3. *Material Characterization*

Tensile Strength Resistance

Tensile strength tests were conducted using a BESMAK (Sincan/Ankara, Turquía) brand universal testing machine with a 5-ton capacity at the SENA Center for Materials and Testing in the regional District Capital. Dogbone-shaped specimens following ASTM D 638 were obtained by injection molding from the various treatments (T1, T2, T3, and T4). This tensile test was performed at a speed of 5 mm/s until the breakpoint.

Hardness

Shore D hardness is used, taking into account that its resistance to indentation will be measured, thus indicating its rigidity and wear resistance. This measurement is essential for comparing the rigidity of different polymers, which helps in the selection of the appropriate material for specific applications.

Hardness is measured by the depth of indentation of the indenter, according to ASTM D 2240 [21]. It is one of the mechanical tests that provides very valuable information about the properties related to resistance to penetration and/or scratching of materials. A CHECK-LINE brand durometer, model MSDD 4AD00 for Shore D hardness, was used. This device is located in the laboratory for composite materials and polymers at the SENA Center for Materials and Testing. The test was carried out at different points on the specimens from the treatments (T1, T2, T3, and T4), where a penetration force was applied, leaving an impression on the material, which allowed for the determination of the corresponding hardness value for each treatment.

Melt Flow Index (MFI)

The MFI is a basic rheological test performed on a polymer to determine its flowability. It is measured in grams per 10 min and is defined as the amount of material (measured in grams) that flows through the orifice of a capillary die in 10 min while maintaining constant standard pressure and temperature according to ASTM D 1238 [22]. The test was conducted using an ATLAS brand plastometer, reference MFI-2, located in the polymer laboratory at the ASTIN National Technical Assistance Center to the industry of SENA regional Valle.

Scanning Electron Microscopy (SEM)

To observe and compare the morphology of the types of microfibers corresponding to the parameters set and shown in Table 1, the scanning electron microscopy (SEM) technique was employed. Micrographs were directly observed in the fracture area of the specimens of each material with treatments T1, T2, T3, and T4 to monitor the behavior of the fibers. Samples were mounted with copper tape to enhance conductivity. A PHENOM XL brand scanning electron microscope was used, which is located at the SENA center for composite materials and polymers and Testing Laboratory.

3. Results

The results obtained in the present research will be described below.

3.1. Electrospinning Process

In the electrospinning process, trials were conducted with the previously established variables, which achieved an average fiber diameter of 7.46 ± 2.5 μm, with fibers as thin as 5.31 μm and as robust as 11.13 μm. A surface roughness on each of the fibers and an excellent structure were also observed, as shown in Figure 3.

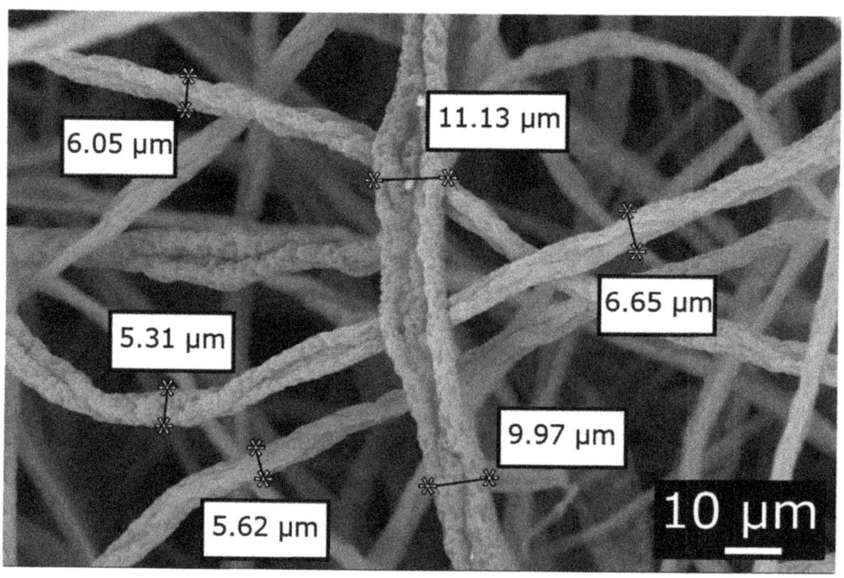

Figure 3. Display of the fiber diameter by SEM of sample T1 2.5 kX.

3.2. Characterization of Developed Materials

Tension Tests

Table 4 presents the tensile test results for the different treatments, where it is evident that the treatments reinforced with MFCA (T1, T2, and T3) exceed the ultimate strength value with values over 11.790 MPa, compared to treatment T4, which does not have reinforcement with cellulose acetate microfibers and shows a value of 11.02 MPa.

Table 4. Experimental design treatments.

Treatment	Ultimate Strength
T1 (LDPE 84.95%, PS 15%, MFCA 0.05%)	12.16 ± 0.8 MPa
T2 (LDPE 84.85%, PS 15%, MFCA 0.15%)	11.79 ± 0.3 MPa
T3 (LDPE 84.7%, PS 15%, MFCA 0.30%)	12.32 ± 0.5 MPa
T4 (LDPE 85%, PS 15%, MFCA 0.00%)	11.02 ± 0.2 MPa

An analysis was also conducted in comparison to the research studies on composite materials with 15% starch carried out by the SENA Materials and Testing Center—GIMES and the Pedagogical and Technological University of Colombia UPTC—DANUM [15,16], where a similar tensile strength behavior is demonstrated. The T1 sample (LDPE 84.95%, PS 15%, MFCA 0.05%) has an average ultimate strength value of 12.16 ± 0.8 MPa, which is a 10.34% increase relative to treatment T4. The T2 treatment (LDPE 84.85%, PS 15%, MFCA 0.15%) has an average ultimate strength value of 11.79 ± 0.3 MPa, which is a 6.98% increase relative to T4, and the T3 treatment (LDPE 84.7%, PS 15%, MFCA 0.30%) has an average ultimate strength value of 12.32 ± 0.5 MPa, an 11.79% increase over T4.

These results suggest that a higher percentage of cellulose microfibers increases the ultimate tensile strength, which could be explained by their high aspect ratio and high tensile strength. These microfibers can provide greater cohesion and strength to the material, resulting in increased tensile resistance, as evidenced in Figure 4.

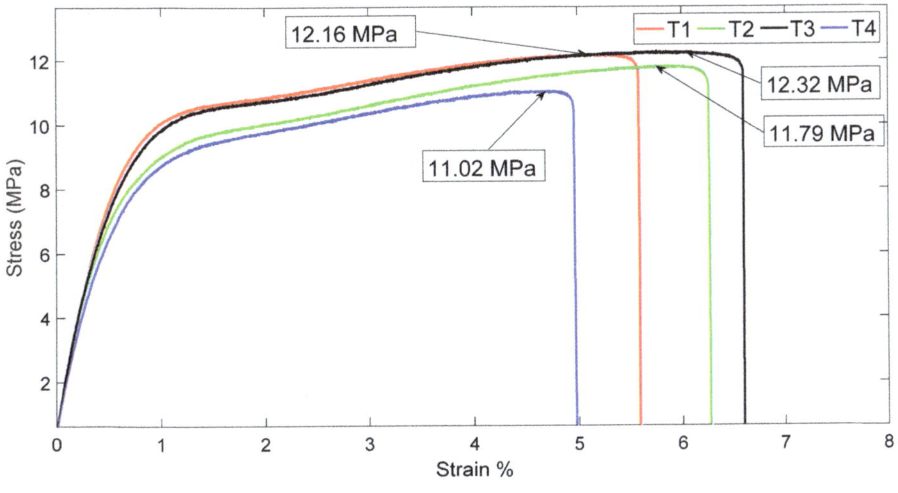

Figure 4. Tensile strength of the treatments made with low-density polyethylene and potato starch.

Hardness Test

For hardness results, the T4 treatment without MFCA reinforcement showed a value of 54.95 Shore D. For the T1 treatment, the average value was 56.25 Shore D, indicating an increase in hardness of 2.36% compared to T4. The T2 treatment (LDPE 84.85%, PS 15%, MFCA 0.15%) had an average hardness value of 54.60 Shore D, a decrease of 0.63% compared to T4. The T3 treatment (LDPE 84.7%, PS 15%, MFCA 0.30%) had an average hardness value of 52.10 Shore D, decreasing by 5.18% compared to T4.

The results, which will be shown in Figure 5, suggest that hardness decreases as the amount of microfibers increases within the developed materials; however, it is noted that the T1 sample has a hardness 1.3 Shore D higher than the T4 sample, with only 0.15% MFCA content.

The findings from the research of authors T. Khan, M. T. B. Sultan, and A. H. Ariffin indicate that a higher percentage of reinforcement increases hardness. However, the hardness values obtained in the current study reveal that only one treatment (T1) shows an increase in hardness compared to T4; yet as more microfiber content is added, the hardness begins to decrease, even falling below the unreinforced T4 treatment [5].

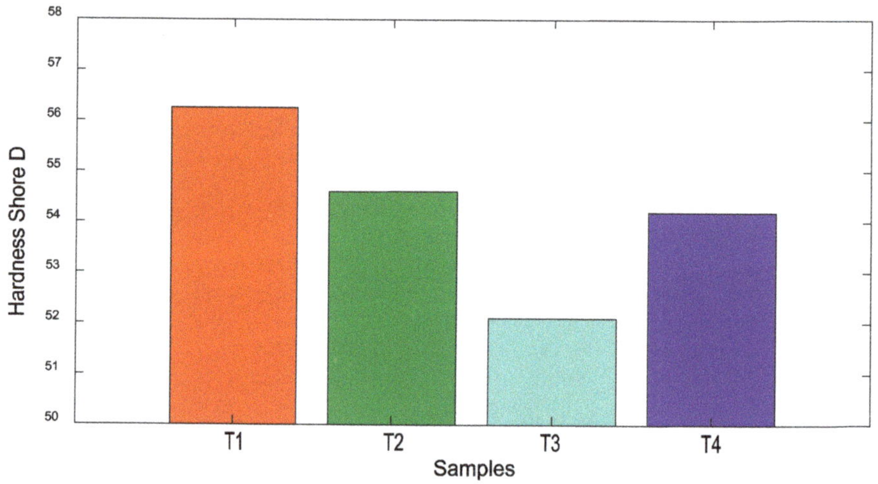

Figure 5. Hardness values of treatments made with potato starch and cellulose microfibers.

Upon examining the hardness values, it can be evidenced that the T2 treatment has a hardness of 54.6, only 0.35 Shore D lower than the T4 treatment, suggesting that these two treatments (T2 and T4) have similar hardness behavior.

Melt Flow Index (MFI)

This test was conducted by measuring the melt flow index of four cuts per treatment and calculating the average value for each sample as recorded in Table 5.

Table 5. Melt flow index values for each treatment.

Treatment Identification	Average Melt Flow Index (g/10 min)
T1	1.69 ± 0.01
T2	1.91 ± 0.13
T3	1.55 ± 0.08
T4	1.95 ± 0.09

The melt flow index for the material used in the test, low-density polyethylene from ECOPETROL, reference POLIFÉN 641, has a value of 1.80 to 2.10 g/10 min. The T4 treatment of this reference had a melt flow index value in pellet form of 1.90 g/10 min.

The average value for the T1 treatment was 1.69 g/10 min and for the T3 treatment was 1.55 g/10 min. These two treatments are below the limit of the T4 treatment, which is 1.95 g/10 min, with T3 being the treatment with the lowest melt flow index.

On the other hand, the T2 treatment had a melt flow index of 1.91 g/10 min, which is within the limits of the technical data sheet for the reference polyethylene (1.80 to 2.10 g/10 min) and very close to the value of T4.

Scanning Electron Microscopy (SEM)

The micrographs were taken from the section of the fractured specimen following the tensile tests of the four developed treatments. The failure area was observed under the scanning electron microscope, from which images were captured and are presented in Figure 6.

In Figure 6a,b, we can observe that in the T1 and T2 treatments, the MFCA creates bonds between the potato starch and the LDPE, leading to agglomerations. However, the MFCA fibers are not detailed, which is attributed to the extrusion process of the three materials that constitute the material for each of the treatments.

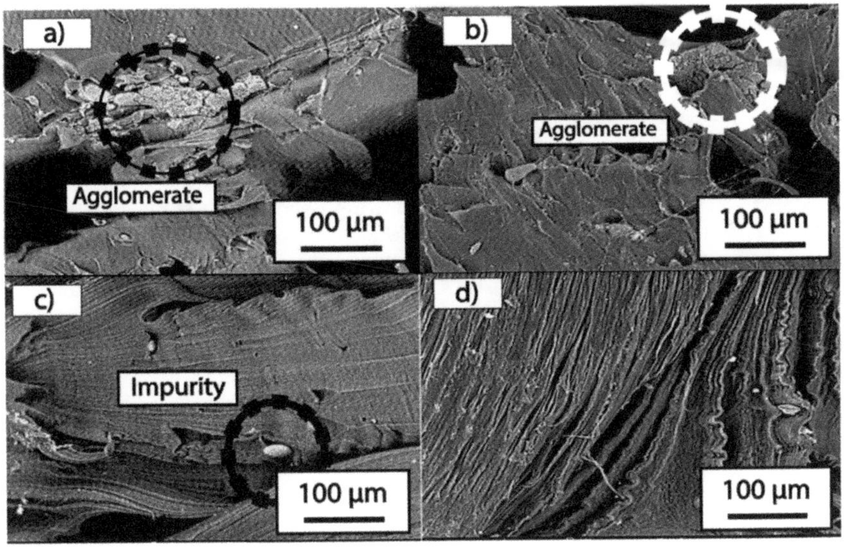

Figure 6. SEM Images of treatments 500× (**a**) T1, (**b**) T2, (**c**) T3, (**d**) T4.

Comparing the T1 and T2 treatments with treatment T3, corresponding to image c, it can be seen that the material presents better homogeneity as no agglomerations are observed in its structure. This is due to the bonds created by the MFCA, which contribute rigidity to the material.

In treatment T4, shown in Figure 6d, a structural orientation is observed, which is due to the stretching applied during the tensile test. Since this T4 treatment does not include any MFCA fiber reinforcement, no agglomerations are visible.

In Figure 6c, the presence of impurities within the polymeric matrix is evident; however, there are no agglomerations as observed in Figure 6a,b, indicated by the black and white circles, respectively.

4. Discussion

Through the electrospinning process and characterization of the microfibers, it was possible to obtain cellulose acetate microfibers (MFCA) with an average diameter of 7.46 ± 2.5 μm, with a range from 5.31 μm to 11.13 μm. The surface roughness observed in these microfibers can contribute to a better adhesion within the polymeric matrix. This characteristic is crucial since a good interface between the fibers and the matrix is essential to improve the mechanical properties of the composite material. In the study carried out by Syed et al. [23] they demonstrated that the electrospinning of cellulose acetate nanofiber reinforced with activated carbon in a mixture of acetone and DMF solvent 2:1 took 3 h to finish with a centrifugation speed of 0.2 mL/h using a voltage of 15 kV. The surface-modified electrospun fiber membrane adheres to pseudo-first-order kinetics and is collected on a flat aluminum foil placed 8 cm away from the needle tip with a removal efficiency of 98.5%.

Tensile tests revealed that the MFCA-reinforced composites (T1, T2, and T3 treatments) showed a significant improvement in ultimate stress compared to the T4 treatment (no reinforcement). Specifically, the T3 treatment (0.30% MFCA) achieved an ultimate stress of 12.32 ± 0.5 MPa, representing an increase of 11.79% over the T4 treatment. This confirms that MFCA reinforcement, due to its high length-to-diameter ratio and tensile strength, improves the mechanical strength of the composite. In the study conducted by Saraswat P, Singh B [24], it shows that 1:3.5 is the optimized plastic to aggregate ratio for LDPE-RCA composite paver blocks with compressive strength of 20 to 35 MPa, suitable for pedestrian or low traffic areas. Indicating that it is crucial to achieve better interfacial bonding, particle stress transfer and ductile behavior, leading to substantial energy absorption before failure.

SEM micrographs provided a detailed view of the internal structure of the composites. Images of treatments T1 and T2 showed agglomerations of MFCA, indicating non-optimal dispersion. However, T3 showed a more homogeneous structure with no visible agglomerations, which could explain the improvement in tensile strength observed in this treatment. The image of treatment T4 confirmed the absence of reinforcement, showing an oriented and stretched structure due to the lack of reinforcing fibers. In the study conducted by Saraswat P, Singh B [24], they show scanning electron micrographs where they reveal the intricate interfacial transition zones (ITZ) within the composite. The well-sorted aggregates, uniform dispersion of LDPE, and robust ITZ formation between LDPE and RCA contribute to better stress transfer, improved orientation, and reduced porosity. Proper interfacial bonding is essential for the mechanical strength and overall durability of the material.

5. Conclusions

The process of electrospinning for the acquisition of cellulose acetate microfibers yielded a fiber diameter standard deviation of 7.46 ± 2.5 µm with a rough surface, successfully standardizing the process for continuous production as per the experimental design proposed for each treatment.

In the mechanical tensile tests, the T3 treatment (LDPE 84.7%, PS 15%, MFCA 0.30%) showed the best result; however, the other two treatments with microfiber reinforcement increase this ultimate strength property of a polymer by 15% compared to a synthetic thermoplastic polymer.

In the hardness tests, the hardness of the T3 treatment (LDPE 84.7%, PS 15%, MFCA 0.30%) is lower than that of a synthetic material, in this case, low-density polyethylene, indicating it is a material that does not easily allow penetration.

In the melt flow index tests, treatments T1 and T3 show a reduction in melt flow index compared to treatment T4, reaching values below 1.7 (g/10 min), while for treatment T2, a melt flow index of 1.91 ± 0.13 (g/10 min) was achieved, close to the T4 treatment of 1.95 ± 0.09 (g/10 min). This indicates that the most suitable treatment for the extrusion process is T2, as it falls within the recommended melt flow index range for LDPE according to its technical data sheet.

The reinforcement with cellulose acetate microfibers improves the mechanical properties of the T4 material, which is solely composed of LDPE and PS. It was observed that a load below 0.3% significantly affects its mechanical properties, referring to various studies used as benchmarks in this research.

This composite material with a thermoplastic matrix reinforced with PS and microfibers (MFCA) makes it a potential candidate for use in industrial processes of this kind for the manufacture of products, providing an alternative to synthetic polymers.

Author Contributions: Conceptualization, P.R.S., A.F.R.-N. and R.V.-G.; methodology, A.F.R.-N., E.Y.G.-P. and R.V.-G.; investigation, R.V.-G.; writing—review and editing, P.R.S.; supervision, and R.V.-G. and E.Y.G.-P.; project administration, E.Y.G.-P.; funding acquisition, and R.V.-G. and E.Y.G.-P. All authors have read and agreed to the published version of the manuscript.

Funding: This research was funded by the Pedagogical and Technological University of Colombia (UPTC), grant number SGI 3746.

Institutional Review Board Statement: Not applicable.

Data Availability Statement: The original contributions presented in the study are included in the article, further inquiries can be directed to the corresponding author.

Conflicts of Interest: The authors declare no conflicts of interest. The funders had no role in the design of the study; in the collection, analyses, or interpretation of data; in the writing of the manuscript; or in the decision to publish the results.

References

1. Laycock, B.; Nikolić, M.; Colwell, J.M.; Gauthier, E.; Halley, P.; Bottle, S.; George, G. Lifetime prediction of biodegradable polymers. *Prog. Polym. Sci.* **2017**, *71*, 144–189. [CrossRef]
2. Langley, J.; Turner, N.; Yoxall, A. Attributes of packaging and influences on waste. *Packag. Technol. Sci.* **2011**, *24*, 161–175. [CrossRef]
3. Kibirkštis, E.; Mayik, V.; Zatserkovna, R.; Vaitasius, K.; Stepanenko, A.; Kandrotaitė-Janutienė, R.; Venytė, I.; Danilovas, P.P. Study of physical and mechanical properties of partially biodegradable LDPE polymeric films and their application for printing and packaging. *Polym. Test.* **2022**, *112*, 107646. [CrossRef]
4. Yuvaraj, G.; Ramesh, M.; Rajeshkumar, L. Carbon and Cellulose-Based Nanoparticle-Reinforced Polymer Nanocomposites: A Critical Review. *Nanomaterials* **2023**, *13*, 1803. [CrossRef]
5. Khan, T.; Sultan, M.T.B.H.; Ariffin, A.H. The challenges of natural fiber in manufacturing, material selection, and technology application: A review. *J. Reinf. Plast. Compos.* **2018**, *37*, 770–779. [CrossRef]
6. Verma, D.; Goh, K.L. Natural fiber-reinforced polymer composites: Application in marine environments. In *Biomass, Biopolymer-Based Materials, and Bioenergy*; Construction, Biomedical, and other Industrial Applications; Woodhead Publishing: Sawston, UK, 2019; pp. 51–73. [CrossRef]
7. Amjad, A.; Rahman, A.A.A.; Awais, H.; Abidin, M.S.Z.; Khan, J. A review investigating the influence of nano-filler addition on the mechanical, thermal and water absorption properties of cellulosic fibre reinforced polymer composite. *J. Ind. Text.* **2022**, *51*, 65S–100S. [CrossRef]
8. Kong, C.; Lee, H.; Park, H. Design and manufacturing of automobile hood using natural composite structure. *Compos. B Eng.* **2016**, *91*, 18–26. [CrossRef]
9. Radoor, S.; Karayil, J.; Rangappa, S.M.; Siengchin, S.; Parameswaranpillai, J. A review on the extraction of pineapple, sisal and abaca fibers and their use as reinforcement in polymer matrix. *eXPRESS Polym. Lett.* **2020**, *14*, 309–335. [CrossRef]
10. Pelissari, F.M.; Andrade-Mahecha, M.M.; Sobral, P.J.D.A.; Menegalli, F.C. Nanocomposites based on banana starch reinforced with cellulose nanofibers isolated from banana peels. *J. Colloid. Interface. Sci.* **2017**, *505*, 154–167. [CrossRef] [PubMed]
11. Rashid, A.B.; Haque, M.; Islam, S.M.M.; Labib, K.M.R.U. Nanotechnology-enhanced fiber-reinforced polymer composites: Recent advancements on processing techniques and applications. *Heliyon* **2024**, *10*, e24692. [CrossRef] [PubMed]
12. Zhang, X.; Wang, Y.; Gao, Z.; Mao, X.; Cheng, J.; Huang, L.; Tang, J. Advances in wound dressing based on electrospinning nanofibers. *J. Appl. Polym. Sci.* **2024**, *141*, e54746. [CrossRef]
13. Ochica, A.F.; Muñoz Prieto, E.D.J.; Vera Graziano, R.; Gómez Pachón, E.Y.; Cerda, A.M.; Rivera Torres, F. Obtention of Cellulose Acetate nanofiberes from Sugar Cane Bagasse. *Cienc. En. Desarro.* **2024**, *8*, 69–77. Available online: http://www.scielo.org.co/scielo.php?script=sci_arttext&pid=S0121-74882017000200069 (accessed on 12 February 2024).
14. Teixeira, E.D.M.; Lotti, C.; Corrêa, A.C.; Teodoro, K.B.R.; Marconcini, J.M.; Mattoso, L.H.C. Thermoplastic corn starch reinforced with cotton cellulose nanofibers. *J. Appl. Polym. Sci.* **2011**, *120*, 2428–2433. [CrossRef]
15. Sandoval, P.R.; Silva, S.M.C. Caracterización de las propiedades mecánicas de un material biodegradable a partir de los procesos de extrusión y peletizado. *Inf. Técnico* **2017**, *81*, 177–180.
16. Sandoval, P.R.; Prieto, E.M.; Pachón, Y.E.G. Obtención y caracterización de un biodegradable a partir de almidón de papa y polietileno de baja densidad por inyección. *Inf. Técnico* **2015**, *79*, 61. Available online: https://www.proquest.com/openview/73641a9e12dc0b897191758e5b7d4677/1?pq-origsite=gscholar&cbl=2043241 (accessed on 12 February 2024).
17. Singh, S.; Gamlath, S.; Wakeling, L. Nutritional aspects of food extrusion: A review. *Int. J. Food Sci. Technol.* **2007**, *42*, 916–929. [CrossRef]
18. Seghar, S.; Asaro, L.; Rolland-Monnet, M.; Hocine, N.A. Thermo-mechanical devulcanization and recycling of rubber industry waste. *Resour. Conserv. Recycl.* **2019**, *144*, 180–186. [CrossRef]
19. Palange, C.; Johns, M.A.; Scurr, D.J.; Phipps, J.S.; Eichhorn, S.J. The effect of the dispersion of microfibrillated cellulose on the mechanical properties of melt-compounded polypropylene–polyethylene copolymer. *Cellulose* **2019**, *26*, 9645–9659. [CrossRef]
20. *ASTM D-638*; Standard Test Method for Tensile Properties of Plastics. ASTM: West Conshohocken, PA, USA, 2022.
21. *ASTM D 2240*; Standard Test Method for Rubber Property—Durometer Hardness. ASTM: West Conshohocken, PA, USA, 2021.
22. *ASTM D 1238*; Standard Test Method for Melt Flow Rates of Thermoplastics by Extrusion Plastometer. ASTM: West Conshohocken, PA, USA, 2013.
23. Syed, N.; Feng, Y.; Mahar, F.K.; Fahad, R.; Abro, Z.A.; Huang, J. Facile synthesize of composite for pollutant removal of lead and methylene blue (MB). *Env. Prog. Sustain. Energy* **2024**, *43*, e14317. [CrossRef]
24. Saraswat, P.; Singh, B. Utilization of recycled concrete aggregates in LDPE-bonded cementless paver blocks. *Constr. Build. Mater.* **2024**, *419*, 135467. [CrossRef]

MDPI AG
Grosspeteranlage 5
4052 Basel
Switzerland
Tel.: +41 61 683 77 34

Polymers Editorial Office
E-mail: polymers@mdpi.com
www.mdpi.com/journal/polymers

www.ingramcontent.com/pod-product-compliance
Lightning Source LLC
Chambersburg PA
CBHW061624210926
51850CB00009B/2426
* 9 7 8 3 7 2 5 8 7 9 5 9 5 *